D0214612

Crops and Carbon

By the same author

Even the Dead are Coming: A Memoir of Sudan

Crops and Carbon

Paying Farmers to Combat Climate Change

Mike Robbins

London • New York

First published 2011
by Earthscan
2 Park Square, Milton Park, Abingdon, Oxon OX14 4RN

Simultaneously published in the USA and Canada
by Earthscan
711 Third Avenue, New York, NY 10017

Earthscan is an imprint of the Taylor & Francis Group, an informa business

British Library Cataloguing in Publication Data
A catalogue record for this book is available from the British Library

Library of Congress Cataloging in Publication Data
Robbins, Michael W.
Crops and carbon : paying farmers to combat climate change / Mike Robbins.
p. cm.
Includes bibliographical references and index.
1. Farmers—Economic conditions. 2. Agriculture—Economic aspects. 3. Sustainable development—Environmental aspects. 4. Agriculture—Environmental aspects. 5. Leadership. I. Title.
HC79.E5R597 2011
363.738'746—dc22
2011004456

ISBN: 978-1-84971-375-7 (hbk)

Typeset in Times New Roman
by JS Typesetting Ltd, Porthcawl, Mid Glamorgan

Printed and bound in Great Britain by the MPG Books Group

Contents

List of Figures, Tables and Boxes

Figures

Tables

Boxes

List of Abbrevations

AFOLU	Agriculture, Forestry and Land Use
AI	artificial insemination
AIJ	activities implemented jointly
BAU	business-as-usual
BBC	British Broadcasting Corporation
BNDES	Banco Nacional do Desenvolvimento
BNF	biological nitrogen fixation
BP	British Petroleum
C&C	contraction and convergence
CA	conservation agriculture
CARB	California Air Resources Board
CBD	Convention on Biodiversity
CCX	Chicago Climate Exchange
CDF	Clean Development Fund
CDM	Clean Development Mechanism
CEIVAP	Comitê para a Integracão da Bacia Hidrografíca do Rio Paraíba do Sul
CER	certified emissions reduction
CGIAR	Consultative Group on International Agricultural Research
COP	Conference of the Parties
CSE	Centre for Science and the Environment
CT	conservation tillage
CTIC	Conservation Tillage Information Center
EB	Executive Board
EMATER	Empresa de Assistência Técnica e Extensão Rural
EMBRAPA	Empresa Brasileira de Pesquisa Agropecuária
ER	emissions reduction
EU ETS	European Union Emissions Trading Scheme
EU	European Union

EUA	European Union Allowance
FACE	free-air CO_2 enrichment
FAO	Food and Agriculture Organization
FFW	food-for-work
FDI	foreign direct investment
FEBRAPDP	Federacão Brasileira de Plantio Direto na Palha
GCM	General Circulation Model
GDP	gross domestic product
GEF	Global Environment Facility
GHG	greenhouse gas
GNP	gross national product
GPG	global public good
GPS	geographic positioning system
IADB	Inter-American Development Bank
IBGE	Instituto Brasileiro de Geografia e Estatística
ICARDA	International Center for Agricultural Research in the Dry Areas
IFOAM	International Federation of Organic Agriculture Movements
IFPRI	International Food Policy Research Institute
IMA	Institute Mineiro de Agropecuária
INCRA	Instituto Nacional de Colonização e Reforma Agrária
INS	inelastic neutron scattering
IPCC	Intergovernmental Panel on Climate Change
JI	Joint Implementation
LAC	Latin America and Caribbean
LDC	least-developed countries
LIBS	laser-induced breakdown spectroscopy
LULUCF	land use, land-use change and forestry
MCA	multi-criteria analysis
MRT	mean residence time
MRV	measurable, reportable and verifiable
MST	Movimento dos Trabalhadores Rurais Sem Terra
NGO	non-governmental organization
NIRS	near infrared reflectance spectroscopy
NPP	net primary production
NT	no-till
ODA	official development assistance

OECD	Organisation for Economic Co-operation and Development
PES	payment for ecosystem (or environmental) services
ppm	parts per million
PRONAF	Programa Nacional de Fortalecimento da Agricultura Familiar
REDD	reduced emissions from deforestation and degradation
RGGI	Regional Greenhouse Gas Initiative
RUSLE	revised universal soil loss equation
SLM	sustainable land management
SOC	soil organic carbon
SOCRATES	Soil Organic Carbon Reserves and Transformations in Agro-Ecosystems
SOM	soil organic matter
SRES	Special Report on Emissions Scenarios
SWC	soil and water conservation
TAR	Third Assessment Report
T-CER	temporary certified emissions reduction
TDR	time domain reflectometry
TERI	Tata Energy Research Institute
UK ETS	United Kingdom Emissions Trading Scheme
UNCED	United Nations Conference on Environment and Development
UNEP	United Nations Environment Programme
UNFCCC	United Nations Framework Convention on Climate Change
USFS	United States Forest Service
VCS	Voluntary Carbon Standard
VER	verified emissions reduction
WMO	World Meteorological Organization
WRI	World Resources Institute
WUE	water-use efficiency
WWF	World Wide Fund for Nature
WOCAT	World Overview of Conservation Approaches and Technologies

Acknowledgements

Many people have assisted or influenced the production of this book, the more so because its origins go back many years. It would be impossible to thank them all properly. I must emphasize that they are not responsible for the book's content and would not necessarily endorse its conclusions.

First I must thank Tim Hardwick of Earthscan and Will Critchley of the Free University of Amsterdam, both of whom have backed the project wholeheartedly. They have also made important and helpful suggestions on the book's content. At the University of East Anglia, John McDonagh and Kate Brown were endlessly supportive during the years this book was taking shape. Other members of the faculty of the School of Development Studies at UEA have also provided inspiration and support; in particular, Piers Blaikie and Michael Stocking have shaped my thinking about rural development and farmers' use of technology. Their views are reflected in this book, but they are not responsible for anything I have misunderstood.

My work at UEA was supported by Britain's Economic and Social Research Council and by the Natural Environment Research Council. I am very grateful for their assistance.

Outside UEA, the roots of this book lie in a conversation about agriculture and climate change with John Ryan, soil scientist at ICARDA in Syria, in the 1990s. And its approach owes much to Michael Zoebisch, whose work and views on farmers and their environment have very strongly influenced mine. His encouragement when I started to think about these issues was invaluable. Many other staff at ICARDA also helped to form the way I think, as did those with whom I worked in the Ministry of Renewable Natural Resources in Bhutan. I also learned much from my colleagues at FAO in 2001 to 2002; in particular, I would like to thank Jose Benites and Theodor Friedrich for discussions on reduced tillage and its implications. I would also like to thank those who gave me time and advice on two trips to Washington, DC, in 2003 and 2005. On both trips, I had long and informative talks with Ian Noble and I am grateful for his time. Especially warm thanks are due to Scott Christensen and Andrea Pape-Christensen for their hospitality on both those trips.

Warm thanks must go to a number of people in Brazil, in particular Bob Boddey and Segundo Urquiaga; there is no question of my work there having been possible without them. Thanks too to Phil and Mercia Chalk for encouragement and hospitality. Eli de Jesus and Leila M. do Valle played a key part in the field survey; Eli also greatly enhanced my understanding of the issues faced by farmers in Minas Gerais. None of these people are in any way responsible for anything I have written, or may have misunderstood, about Brazil. I would also like to thank everyone, particularly the farmers, extension staff and other officials, who gave up their time to meet me and discuss their work. As this book goes to press, the State of Rio de Janeiro is coping with the aftermath of appalling floods that have taken hundreds of lives; I would like to offer my sympathies, and wish the state well for its future.

Last but not least, one's friends are the environment in which one's work is done. I cannot, I suppose, list them all; but Anna Allen, Patricia Almaguer-Kalixto, Emily Boyd, Dan Coppard, Marta Einarsdóttir, Ian Fuller, Lawrie Hallett, Joe Hill, Sylvie Koestlé, and Neil Monk, among many others, have been especially supportive at various times. And special thanks must go to Hazel Marsh in Norwich and Lisa Sutton in New York, for their warmth and humour and for keeping me sane as this project neared its end.

Introduction

It is reasonable to assume that the dry margins of semiarid lands contracted or expanded according to climatic variations. But to propose that civilization was chased from Babylon to London by a creeping drought is not reasonable. The evidence for climate as a destroyer of civilization is vague or nonexistent; the evidence of soil erosion is there for all to see.

(Carter and Dale, *Topsoil and Civilization*,1955)

The climate appears to be changing due to human activity. To stop or reverse this, there are now national and international systems of incentives for activities to mitigate emissions of greenhouse gases (GHGs). One such activity is sequestration of carbon dioxide – CO_2 – into sinks. The best-known carbon sinks are forests, but agricultural land is also a sink, and it is one that can be enhanced or degraded, depending on the way the land is farmed. Farmers in the developing world might benefit from this, either by receiving direct payments for the carbon removals, or through development funding leveraged by the environmental benefits. At least, that is the theory. The objective of this book is to examine the practical possibilities.

There is a clear theoretical rationale for considering agricultural sinks as a mitigation strategy. CO_2 emissions are fuelling climate change. Sinks are carbon pools, the contents of which are *not* adding to the atmospheric carbon pool because they are held elsewhere. The size and nature of sinks is highly variable; forests are the best-known but much carbon is also held in aquatic ecosystems, transported there by soil erosion, while the oceans absorb a considerable amount of carbon dioxide – and undersea formation of limestone is composed chiefly of carbon. Other carbon deposits include fossil fuels such as oil and coal, formed over millennia from plant material. The stability of these sinks clearly varies; fossil fuels are stable until disturbed by human agency, but forests are subject both to this and to natural fires.

The soil organic carbon pool is also subject to disturbance both by human activity and natural processes, and its size is significant. Estimates vary, but one suggested figure is 1550Pg[1] (Follett, 2001, pp78–89).This is roughly twice the atmospheric carbon pool, which is 770Pg (Lal, 2002, p353). Sombroek et al (1993, quoted in UNCCD, 2008, p3) suggest that the soil carbon sink in the top 1m of soil is 1.5 times the total for standing biomass. Soil carbon can be converted to CO_2 through land-use change, ploughing and erosive processes that are mainly connected to agriculture. Agriculture has therefore caused huge losses to the atmospheric pool ever since it began in settled form about 8000 years ago. How much, is arguable; Lal (2004a, p1623) reports that estimates vary from 44 to 537Pg, but that they typically range from 55 to 78Pg.

However, there is widespread acknowledgement that agriculture can, at least in theory, recover much of this. Lal and Bruce (1999, p178), who estimate the historic loss at 55Pg C, also believe that as much as 75 per cent might be recoverable through the sequestration of CO_2 into the soil – that is, its conversion into organic carbon, contained in plant material, through photosynthesis. FAO (2001a, p61) suggests 23–44Pg C in agricultural soils over the next 50 years; this implies that agriculture can be of comparable significance to forests, the global sequestration potential of which has been estimated at 60–87Pg C (ibid.). This sequestered carbon can to some extent be converted into soil organic matter (SOM) through (for example) the decomposition of root systems and the incorporation of crop residues. This in turn can have positive inputs to productivity, soil health and the sustainability of agriculture.

In 1998 nearly 100 North American scientists and agriculture-industry figures met at St Michaels, Maryland, to discuss the potential. In his summary of the workshop, Rosenberg reports it found that:

> *[R]eductions in atmospheric carbon content can be achieved by large-scale application of tried-and-true land management practices such as reduced tillage; increased use of rotational crops such as alfalfa, clover and soybeans; and by an efficient return of animal wastes to the soil. Forests and grasslands afford additional capacity for carbon sequestration when established on former croplands. Programs to further soil carbon sequestration will provide ancillary benefits including improvements in soil fertility, water holding capacity, and tilth, and reductions in wind and water erosion.* (Rosenberg et al, 1999, p1)

Although much of the potential lies in the developed world, increasing SOM could have particular significance in regions where, for economic reasons, farming must be pursued in fragile ecosystems; many of these are in poorer countries where low organic matter limits productivity and resistance to erosion. So sequestration of carbon in agriculture can both slow climate change, and benefit the rural poor. Meanwhile, the development of carbon-trading instruments means that the carbon may have cash value, and this might be used to fund initiatives to increase soil carbon. Despite this, agricultural sinks are currently excluded from the Clean Development Mechanism (CDM) of the Kyoto Protocol, and it is by no means assured that they will be eligible for any Kyoto Phase II, should such an agreement come into force after the current agreement expires in 2012.

This is despite agricultural sinks appearing to be a true 'win-win' strategy with positive implications not only for climate-change mitigation and poverty alleviation, but also adaptation to climate change. Soils with more organic matter will better resist the erosive processes that might be increased in years to come – for example, greater temporal concentration of rainfall. Such soils will also have better water-infiltration capacity, which will be important for the same reason.

However, there are pitfalls. Some are fundamental. Is carbon really a commodity that can be priced, bought and sold in any meaningful sense? Our understanding of value in a market society would suggest not. If this is the case, then the 'carbon market' is a bureaucratic fiction that farmers would do well to avoid. And even if one accepts that carbon does have value, how does one price it?

There are also many questions specific to agriculture. Some of the agronomic practices required to sequester carbon are those already advocated for soil conservation; attempts to support this have had a difficult history in agricultural development. Linking these practices to carbon funding, as well as soil conservation, would not automatically eliminate these difficulties. Even if it did, there would still be major methodological constraints to linking agricultural carbon to the market. Not least would be additionality – that is, would the carbon have been sequestered anyway without the project? Proving that it would not is especially hard when the measures taken have other benefits.

Monitoring and verification of agricultural sinks is also challenging. To begin with, to prove that a carbon sink has been enhanced, it is necessary to know not just how large it was before the project, but how much carbon might have been sequestered or lost without intervention. This is called the baseline, and it is fundamental to any project, in any sector, that seeks to monetize emissions reductions or GHG removals.

Assuming that a baseline has been set, the amount of carbon being sequestered, or not emitted, from the farm must be monitored. Above-ground carbon – trees and other vegetation – can be quantified, although there are challenges. The soil carbon sink is more difficult. Tools to measure soil carbon have been around for a long time, but they measure a sample taken from a given point, and the amount may vary widely even within a single field.

A major methodological challenge is permanence. This applies to all land-use projects; how can one say a forest will never be cut down? The same applies to agricultural land; a farmer may agree to apply minimum tillage in order to increase the carbon in the soil, but can plough again very easily if there is some incentive to do so. Sinks have fared badly in negotiations partly for this reason.

Who is going to invest in carbon credits from agriculture? It will not be the cheapest way to generate them. And will there be a market? The Kyoto Protocol expires in 2012; it might not be replaced, and has provided few funding mechanisms for agriculture anyway. What are the alternatives?

This book examines these questions. It starts by explaining climate change and what it will do, especially to agriculture. Next it explains what agriculture can do to mitigate this process, and how. Other chapters review the carbon market and agriculture's place in it, and examine the methodological constraints mentioned above. Finally, the book examines the potential for agricultural sequestration in one biome – the Atlantic Forest region of Brazil. This includes how the above- and below-ground carbon stocks in the region become depleted, and what the farmers think the constraints might be to introducing carbon-friendly farming practices. This was done with a farmer survey. The responses, and the views expressed by the farmers, have helped to illuminate what some of the deeper, less visible constraints to 'carbon farming' might be.

Several themes emerge from this book. They include the need to fit carbon mitigation into both global and local contexts; it cannot be thought of in isolation. They also include the need to confront complexity; there is nothing simple about emissions mitigation, especially in agriculture. Not least of the complexities is the relationship between mitigation of

CO_2 emissions, and the other greenhouse gases closely associated with agriculture, nitrous oxide (N_2O) and methane (CH_4).

But one of the most important themes is scientific uncertainty. The unknowns around climate change itself are common knowledge. The Intergovernmental Panel on Climate Change (IPCC) has always clearly acknowledged them in its successive assessment reports. Climate science operates at the frontiers of knowledge and is essentially inductive; it must try to construct the future from the facts it has so far.

What is perhaps less well understood, is that there are sometimes specialized gaps in the knowledge regarding climate change that are also important, and *could* be filled. Data concerning (for example) the relationship between carbon emissions and soil erosion, or the agronomic properties of biochar, is not available because until recently nobody knew that they would be needed. It seems unlikely that this phenomenon is unique to agriculture.

The author hopes that this book will demonstrate the potential and, despite the challenges, feasibility of agricultural sinks in the developing world. But another message, almost as pressing, may be the importance of the biophysical sciences at every level in the drive to mitigate climate change.

Finally, a word of warning: climate science, and climate negotiations, are constantly moving on. The author has tried to ensure that the contents of this book are up to date at the time of going to press (February 2011); inevitably, however, there will be relevant events, and new papers published, by the time it appears.

Note

1 1Pg (petagram) C is equivalent to 10^{15}g – that is, 1000 million metric tonnes; this can also be expressed as gigatonnes (Gt). Also used is a Tg, or teragram, of C (10^{12}g), 1,000,000 tonnes. To reduce confusion, these measurements will generally be quoted in the same way as they are at source.

1

Climate Change: Implications for Agriculture

The movements of the air and the waters, the extent of the seas, the elevation and the form of the surface, the effects of human industry and all the accidental changes to the terrestrial surface modify the temperatures in each climate. (Jean-Baptiste Fourier, *Memoire sur les temperatures du globe terrestre et des espaces planetaires*,1827, trans. W. M. Connolley)

The hypothesis behind this book is that agriculture could, with improved management practices, absorb a significant amount of carbon dioxide (CO_2) as soil carbon, and that farmers in the developing world might in some way benefit from its value as a mitigation strategy for climate change. This presupposes that climate change is taking place and is due in part to CO_2, that agricultural soils could hold more of it than they do, and that there could be a climatically significant removal of CO_2 if they did. As these are large assumptions, this chapter and the next set out the background.

First, climate change is introduced, as some background may be needed for readers whose background is mainly in agriculture. It is a brief introduction; there are plenty of books available for those who wish to consider it further. For the same reason, the probable consequences of climate change are described mainly in their agricultural context. Those who wish to look at its broader impacts could start with the assessment reports of the IPCC and the wide range of papers that they quote.

Anthropogenic Climate Change

The notion that human activity changes the climate is not new. In the Middle Ages, it was thought magic was the human agency, and that burning witches would alleviate flood or drought. This persisted. Baliunas (2004) argues that the Little Ice Age of roughly 1500–1750, following as it did a warmer medieval climate, provoked a wave of witch hunts; and several thousand executions appear to have taken place in Bamberg, Würzburg, Electorate Mainz and Westphalia in 1626 as a response to a severe late frost that destroyed both grain and vine in the spring of that year, prompting popular wrath at witchcraft.

However, anthropogenic climate change was probably first considered in the modern sense in the early 19th century when French scientist Jean-Baptiste Fourier put forward the view that gases in the atmosphere affected the surface temperature of the earth through a 'greenhouse effect'. Fourier was a polymath whose activities included physics, mathematics, revolutionary politics and governing Grenoble; he constructed his theory not long before his death in 1830, and did little to link this effect to human activity. It would be left to Irish-born researcher John Tyndall, later in the century, to identify the causative gases and to include among them carbon dioxide. Tyndall – a polymath like Fourier, with an interest in travel, mountaineering and literature – would also be the first to demonstrate the 'greenhouse' phenomenon in the laboratory, where he used the first ratio spectrophotometer to establish the different absorptive powers between a number of gases, including CO_2 (Fleming, 1998, p1).

The link to human activity was first clearly expounded by the Nobel prizewinner Svante Arrhenius in the early 20th century. His Nobel Prize (in 1903) was for electrochemistry, but he was also well aware of Tyndall's work, and eight years earlier he had calculated the probable influence on the climate of a rise or fall in CO_2 concentrations – the first attempt to quantify it as such; a 2.5 times or threefold increase would, he thought, warm the Arctic regions by 8–9°C (Earth Observatory, 2005a, p1). Although both Fournier and Tyndall understood that some 'greenhouse gases' were emitted by human activity, Arrhenius seems to have been the first to argue that this could really change the climate, writing in 1904 that 'the slight percentage of carbonic acid [CO_2] in the atmosphere may, by the advances of industry, be changed to a noticeable degree in the course of a few centuries' (ibid.). This left him untroubled. It would, he suggested, produce more food; and, being Swedish, he might have welcomed a warmer winter.

The perception that anthropogenic climate change might be a threat is relatively recent. In 1958 the Scripps Institution of Oceanography began to monitor atmospheric CO_2 at the Mauna Loa observatory in Hawaii (Earth Observatory, 2005b, p1); the previous year, Roger Revelle and Hans Suess had demonstrated the anthropogenic effect on the atmospheric carbon pool in a seminal article in *Tellus* (Earth Observatory, 2005b, p1; Scripps Institution, 2005, p1). In 1965, Revelle brought the potential dangers to public and government attention as a member of the President's Science Advisory Committee Panel on Environmental Pollution. Revelle is sometimes regarded as one of the key figures in climate-change research; receiving the National Medal for Science in 1990 (from the first President George Bush), he said, 'I got it for being the grandfather of the greenhouse effect' (Scripps Institution, 2005).

Today, thanks to techniques such as the analysis of ice cores, it is possible to measure historic CO_2 levels and thus tie the increase in atmospheric CO_2 more or less to the Industrial Revolution in the middle of the 18th century. The Intergovernmental Panel on Climate Change (IPCC) states that the the pre-industrial atmospheric CO_2 concentration levels were 280 ± 10 parts per million (ppm) in 1750, and had been at around that level for several thousand years; they then rose to 367ppm in 1999 and 379ppm in 2005 (IPCC, 2001a, p185; 2007a, p2). It adds: 'The atmospheric concentration of carbon dioxide in 2005 exceeds by far the natural range over the last 650,000 years (180 to 300ppm) as determined from ice cores' (IPCC, 2007a, p2). The process has increased recently. According to Dlugokencky (2010, p41–42), of roughly 335 billion tons of carbon that has been emitted into the atmosphere by human activity since 1750, about half has been emitted since the mid-1970s. Emissions rates increased by over 3 per cent from 2000 to 2004 (Raupach et al, 2007, quoted in Dlugokencky, 2010, p42).

However, it is one thing to identify and measure anthropogenic greenhouse emissions, but another to quantify the results. As stated earlier, Tyndall established the differential conductivity of the gases present in 1859, and Arrhenius subsequently tried to quantify the effect on climate, but projection of climatic effects is subject to a myriad of variables and unknowns. For example, atmospheric carbon is increasing at only half the rate of emission from fossil fuels, due to terrestrial uptake (Raupach et al, 2007, quoted in Dlugokencky, 2010, p41). But no one is sure exactly where this remaining 50 per cent is going; it is being absorbed by forests, soils and oceans, but these processes cannot yet be accurately quantified – giving rise to the notorious 'missing sink' (see for example Houghton et al,

1998). Moreover there are substantial year-on-year variations; the rate of increase was 3.3 ± 0.1Pg/yr C during 1980 to 1989 and 3.2 ± 0.1Pg/yr C during 1990 to 1999 (Raupach et al, 2007, quoted in Dlugokencky, 2010, p41), but 1992's figure was 1.9Pg and 1998's was 6Pg, thanks mainly to variation in terrestrial uptake itself induced by volcanic activity and/or climatic variation.

Not least of the factors here could be terrestrial feedbacks to climate change from processes that it has itself provoked. This could include emissions from sinks such as agricultural land and in particular, peat bogs, where higher temperatures could induce faster mineralization of soil carbon (as could fires, of course; the extreme heat and peat fires seen in European Russia in 2010 must be a cause of concern in this respect). This could be coupled with an increased emission of methane, and this may be happening; methane is a powerful greenhouse gas, and Dlugokencky (2010, p42) reports that after a decade of near-zero growth, atmospheric methane increased globally in 2007 and 2008, mostly from emissions in the Arctic in 2007 and the tropics in 2007 and 2008. Dlugokencky comments that: 'Likely drivers for increased emissions are anomalously high temperatures and precipitation in wetland regions, particularly in the Arctic during 2007.'

These factors must be built into any future projection, although they remain highly uncertain. Moreover these projections – the product of General Circulation Models, or GCMs – do not always agree in their assessment of either past or future climatic trends. So GCMs, on which forecasts of climate change depend, could be presented as informed speculation. Whether or not they are viewed as such may depend on one's assessment of induction as a tool of enquiry. But it is known that for much of the 20th century, the global surface temperature increased at about 0.15°C per decade. The Fourth Assessment Report of the Intergovernmental Panel on Climate Change in 2007 stated that 11 of the 12 years from 1995 to 2006 had been amongst the 12 warmest years since 1850, and that the 100-year linear trend to 2000 had been an annual increase in global surface temperatures of 0.74°C. This was a little higher than supposed at the time of the Third Assessment Report, which put it at 0.6°C (IPCC, 2007a, p5). More recently, in *State of the Climate in 2009*, Menne and Kennedy (in Arndt et al, 2010, S24) state that the decade 2000–2009 was the warmest of the last five decades: 'Decades1960–69 and 1970–79, were cooler than 2000–09 by about 0.6°C, with 1980–89 cooler by about 0.35°C and 1990–99 cooler by 0.2°C.' Will this continue?

The IPCC has put together a series of projections of future trends. Finalized just too late for the Third Assessment Report, they were summarized in it in provisional form and then published shortly afterwards as the Special Report on Emissions Scenarios, or SRES (IPCC, 2001b), which is extensively quoted in the Summary for Policymakers of the Fourth Assessment Report (IPCC, 2007a).

The SRES report seeks to project emissions scenarios according to different economic and social trends, and in so doing it indicates how complex climate forecasting would be even if GCMs were known to be precise. The report produced no less than 40 scenarios, eventually amalgamated into just four basic 'marker' scenarios labelled A1, A2, B1 and B2. A1 assumes rapid economic growth, a population peak in the mid-21st century, and reductions in income disparities. Given this background, it subdivides into three possible fossil-fuel use scenarios: intensive, use of alternatives, or a balance across several sources. The other three groups also assume different combinations of economic growth, shifts to service economies, and balances of sustainable and other energy sources.

The resulting emissions projections cover a very broad range even when run through only one or two GCMs (a source of variability in themselves). For a start, the Third Assessment Report projects carbon emissions by 2100 at anything from 770 to 2540Gt C. The IPCC states that variations in 'climate sensitivity and ocean and terrestrial model responses' add at least −10 per cent to +30 per cent uncertainty to the level of atmospheric carbon that this would create. Once uncertainties regarding terrestrial feedbacks have been incorporated, this would indicate atmospheric CO_2 concentrations of anything between 490 and 1260ppm, or between 75 per cent and 350 per cent above those in 1750, at the dawn of the industrial age. Thus the global average temperature increase was projected at anything between 1.4 and 5.8°C above 1990 in the Third Assessment Report.

The Fourth Assessment Report, published in 2007, is a little more pessimistic, suggesting a likely range of 1.8 to 6.4°C. It suggests that this is more accurate as it is based on 'a larger number of climate models of increasing complexity and realism' (IPCC, 2007a, p13), and also now incorporates some information on feedbacks in the climate cycle. This is important because, as described above, rising temperatures could (for example) increase emissions from peat bogs or the rate of mineralization of soil carbon, which could in turn accelerate the rise in temperature. It should be noted that the upper range is predicated on the A1F1 scenario, that is 'a future of very rapid economic growth' which is fossil-fuel intensive, albeit with more efficient technology. However, the lower, B1, scenario posits a less globalized world and 'intermediate' economic growth, with emphasis on local solutions. Neither seems entirely likely; the A1 scenario is perhaps more probable, but on the other hand all scenarios omit any type of international agreement to limit emissions (op. cit., p18); so on balance we might expect something halfway between the upper and lower estimates.

However, none of this means that climate change is intrinsically bad; as has been said, Arrhenius did not see it as such. So it is worth looking at the probable impacts of climate change; that they are worth mitigating is also an underlying assumption of this book, and that should be examined.

What Climate Change Will Do

Predicting what such rises in temperature would mean in practice is yet more complex. In general, precipitation is likely to rise in higher latitudes all the year round, with increases in winter in tropical Africa and in summer in South-east Asia. In Australia, Southern Africa and Central America, however, winter rainfall is expected to decrease. In fact, the Fourth Assessment Report states that the 20th century also saw land-surface rainfall in the subtropics decrease by about 0.3 per cent a decade.

The picture is less clear for the tropics, where it has increased by roughly the same amount, but not in the last few decades. However, there is some evidence that it *has* increased over the oceans. Over the last century, there has been a 'significant' rise in precipitation over the eastern parts of North and South America, northern Europe and northern and central Asia, but a decline over the Sahel, Mediterranean, southern Africa and parts of southern Asia (IPCC, 2007a, p7).

Intensity of precipitation is seen as increasing almost everywhere, as is the frequency of extreme precipitation events (IPCC, 2001b, pp71–72). There is already evidence of this; Levinson et al (2010, p31) report that 2009 saw the continuation of 'an extended period of above average global [land] precipitation covering the past five years ... and much of the period since 1996'. This will have some positive implications (along with some negative ones, such as flooding – and more soil erosion, especially in fragile, semi-arid environments where rainfall is already very concentrated in time and space). Generally, however, global food production is likely to suffer. True, the Fourth Assessment Report suggests that there will be 'small beneficial impacts on the main cereal crops' with a temperature increase of 1–3°C, but that further warming would have increasingly negative impacts (the IPCC assigns medium to low confidence to this prediction). Moreover in lower latitudes, even small temperature increases are expected to reduce production of the main cereals (IPCC, 2007b, p285). Given the global distribution of cereal production, it is argued that it would see a net decline if the rise in temperature exceeded 3°C.

Up to that level, however, there is a chance that gains in food production in higher latitudes could offset losses elsewhere in terms of global food production. But the gains would not, in the main, be in the regions where the poorer and less food-secure live. As the Third Assessment Report put it, there would be 'economic impacts on vulnerable populations such as smallholder producers and poor urban consumers. [Recent studies find] that climate change would lower incomes of the vulnerable populations and increase the absolute number of people at risk of hunger, though this is uncertain and requires further research' (IPCC, 2001c, p11).

The rising level of malnutrition is also seen as causing impaired child development and 'decreased adult activity' (IPCC, 2001c, p12) – while the geographical reach of malaria and dengue is expected to increase. On one projection, an additional 300 million people could be at risk from falciparum (cerebral) malaria by 2080, and perhaps 150 million from vivax malaria (Martens et al, 1999, pS102). The IPCC also adds that: 'Extensive experience makes clear that any increase in flooding will increase the risk of drowning' (IPCC, 2001c, p11). This rather deadpan statement has more force in the aftermath of the 2010–2011 floods in Pakistan and elsewhere.

Estimating to what extent crops would be affected is extremely complex, not least because changing climatic conditions could be important at highly specific stages in crop development. As one working group to the Fourth Assessment Report (IPCC, 2007b, p277) points out, 'yield damage estimates from coupled crop–climate models need to have a temporal resolution of no more than a few days and to include detailed phenology'. It may also be hard to separate the effects of climate change from loss of productivity due to other degradation resulting from human activity, such as soil erosion and soil salinity – effects that will themselves increase the vulnerability of production systems to climate change (IPCC, 2007b, pp277–278). A particularly challenging variable is how the CO_2 'fertilization effect' will act on both food production and terrestrial CO_2 uptake, and this is reviewed briefly later in this chapter.

The uncertainty is exemplified by the trends in African and Latin American maize production projected by Jones and Thornton (2003). They foresee three basic types of response to climate change. One, to be sure, is grim – maize yields collapse and human populations are displaced as a result (op. cit., p55). However, great care is needed with these projections. Although the picture for lower latitudes is generally poor, some areas may benefit; while in other areas, such as eastern Brazil, there will be yield decreases but within a scale that can be dealt with through crop breeding and improved agronomy.

The IPCC itself made an unintentionally misleading statement in the Fourth Assessment Report's Synthesis for Policymakers when it stated that some African countries could see drops in yields from rainfed agriculture of up to 50 per cent in the 2000–2020 period. The statement was based on a misinterpretation of sources that were legitimate in themselves, but actually only gave that figure for that period with reference to Morocco (Carr et al, 2010, p21). In fact, crop yields in some parts of Africa may benefit, at least in the near future. Jones and Thornton see the Ethiopian highlands around Addis Ababa as experiencing anything up to 100 per cent yield increases. Doherty et al (2010) looked at the prospects for East Africa and found a likely increase in crop yields in some areas. They used the SRES A2 scenario, which suggests quite a large rise in emissions through the century; this would have a greater climatic impact but would also lead to more net primary production (NPP) – that is, production of plant material – due to the CO_2 fertilization effect, along with some other factors. 'Increased rainfall, river runoff and fresh water availability alongside enhanced NPP may be tentatively expected to improve conditions for agriculture, provided temperature is not a limiting factor,' they state (op. cit., p22). They also foresee a small drop in soil carbon after 2050 (op. cit., p10).

This uncertainty is naturally reflected in the forecast of economic impacts. The IPCC's Second Assessment Report suggested that a doubling of CO_2 would cost about 1.5 per cent of GNP, but this was later thought to have been pessimistic (Mendelsohn and Williams, 2004, p316). But counting the cost of environmental change is a contentious business; and in any case, costing it on a global basis is not always helpful. There is a similarity here with land degradation; as Scherr (2001) has explained, that could cut production by up to 17 per cent by 2030, but the world food system would adjust for these losses. However, she points out that the impacts *would* be felt severely at a local level, where they would be worst for the poorest. Climate change is no different in this respect. And there will be plenty of other variables; apparently similar agroecological zones would have significant differences in vulnerability depending on the fertility of their soils and their ability to retain moisture, the propensity of the local farmers to maintain cover crops, and their access to crop varieties bred for moisture stress (or simply to adaptable landraces and wild relatives). These complexities underpin a growing research effort on vulnerability and adaptation to climate change and the factors that influence them. However, it is clear that one consequence of climate change could be hunger. Parry et al (1999, S51) suggest that there could be between 55 and 65 million extra people at risk of hunger in Africa by 2080 because of climate change (although this is not a huge percentage increase; FAO already states that 925 million people are 'chronically hungry').

Uncertainty: The CO_2 Fertilization Effect

However, the uncertainties are as enormous for agriculture as they are for anything else. The CO_2 fertilization effect, mentioned earlier, is a case in point. Elevated levels of CO_2 cause partial closing of plant leaf stomata, reducing transpiration for a given area and raising leaf temperature (Wittwer, 1995, p2). More of the CO_2 taken in by the plant is converted to plant material. This is important for two reasons. First, it implies that, far from damaging agriculture by changing the climate, increasing CO_2 levels may make it more profitable – not least because reducing transpiration also improves water-use efficiency (WUE), potentially vital in the arid areas that will be most vulnerable to climate change. Second, it may change the balance between photosynthesis and autotrophic respiration – that is to say, plants may convert a greater percentage of their CO_2 intake into organic matter, instead of re-releasing it. This increase in NPP would thus be mitigating climate change through reduction of the atmospheric CO_2 pool.[1]

Some have gone so far as to argue that this could even offset the negative effects of climate change. One such is Sylvan Wittwer, a horticulturalist and former head of the Agricultural Experiment Station at Michigan State University. He has argued that the CO_2 fertilization effect will mitigate any effect climate change has on food supplies. In his 1995 book

Food, Climate and Carbon Dioxide, Wittwer stated that the 'positive, beneficial and directly measurable' effects of increased atmospheric CO_2 are 'glossed over lightly or submerged in a series of catastrophic issues. These are associated with a widely promulgated global warming, which has been greatly exaggerated … and which may not even exist' (Wittwer, 1995, p2).

This book is now 16 years old, but Wittwer has continued to advise the Center for the Study of Carbon Dioxide and Global Change, the mission statement of which is that it 'attempts to separate reality from rhetoric in the emotionally-charged debate that swirls around the subject of carbon dioxide and global change. In addition, to help students and teachers gain greater insight into the biological aspects of this phenomenon, the Center maintains on-line instructions on how to conduct CO_2 enrichment and depletion experiments.' It publishes an online bulletin, *CO_2 Science*. The central aim seems to be to argue that rising levels of CO_2 will be beneficial to humanity. (As well, perhaps, as promoting climate scepticism; the site includes a feature called 'Medieval Warm Period of the Week'.)

That increased levels of CO_2 will increase productivity in some cases is not in doubt, so it is worth considering this argument. The IPCC also acknowledges that the resulting increase in NPP will have climate feedback effects (IPCC, 2001a, p51); although not considered in the Second Assessment Report, they were incorporated in the projections for the Third (IPCC, 2001a, p186), and were also considered by the working group on food, fibre and forest products for the Fourth Assessment Report (IPCC, 2007b, pp282–283, 290), and by the group on ecosystems and their services (IPCC, 2007c, p220).

However, the extent to which NPP increases, the probability of increased food production, and whether there is a negative feedback on climate change, are dependent on several variables. Improvements in WUE will not be significant in non-water-stressed environments, where it is not an issue (Dubrovsky et al, 2000, p2). It may also be unhelpful in environments that are *very* water-stressed, as these are already marginal for agriculture and an increase in temperature may mean that soil moisture ceases altogether to be adequate for crop production. The same applies to grassland species; the Fourth Assessment Report's working group on ecosystems and their services points out that apparent responses to CO_2 enrichment are occurring in a given moisture regime (IPCC, 2007c, p220). That regime may of course change with rising temperatures; and it is pointless using water better if there isn't any.

Moisture apart, increased temperature in its own right would also offset gains to some extent, especially in heat-stressed environments. Planting dates can (and probably often will) be altered to allow for rising temperatures, but this will be constrained by photoperiod and by less predictable rainfall. Indeed Canadell et al (2007, p372) suggest that the gains will be offset by heterotrophic respiration. Last but not least, species with different photosynthetic pathways – that is to say, ways of fixing CO_2 and turning it into plant material – will react differently; some types usually respond to elevated CO_2, but others are already more efficient in their processing of CO_2, and show less or no response (IPCC, 2001a, p195). The first type includes most plants in colder climates, wheat, rice and most agricultural crops. However, the second includes maize and sugar cane; also tropical and some temperate grasses. This could imply positive feedback effects to climate change, as loss of carbon through decomposition of organic matter is likely to be much faster in tropical climates. In that situation, organic matter may not accumulate any more quickly but may break down faster as the temperature rises. As organic matter content is important for productivity, this will in some cases reduce NPP, further reducing the accumulation of soil organic matter. (In very hot and/or arid areas, this may set off a spiral that would not only reduce food production but would convert large regions from carbon sinks into carbon sources; but this is discussed later.)

Moreover, 'real-world' response of plants to elevated CO_2 is imperfectly understood. The standard technique for measuring response has been to use chamber fumigation, but as Morgan et al (2005, p1857) report, using free-air CO_2 enrichment (FACE) simulates elevated CO_2 levels without otherwise altering the microclimate. The authors report previous FACE trials that suggest wheat and rice had a response to elevated CO_2 levels rather lower than that predicted by chamber studies; moreover their own FACE trials with soybean showed a substantially lower increase in yield than would have been expected on the basis of previous chamber studies. They argue that future projections of global food supply increases may be optimistic (Morgan et al, 2005, p1864). It should be noted, however, that other researchers have been less sure that responses to CO_2 are overestimated (IPCC, 2007b, p282).

Parry et al (1999, S56) list further uncertainties in predicting yield response (and presumably NPP too) from trials: weeds, diseases and insect pests are assumed to be controlled; there are no soil problems (for example acidity or salinity); and no extreme weather events. Parry et al (2004, p66) also modelled the yield of the world's major food crops against the IPCC's SRES scenarios, and suggested yield declines of 0–5 per cent. However, they warn that these yield declines are in part dependent on the CO_2 fertilization effect being fully realized – e.g., it is implied that they could be bigger: 'The realisation of … potential beneficial effects of CO_2 in the field remain uncertain due primarily to potential, yet still undocumented, interactions with nutrients, water, weeds, pests, and other stresses' (Parry et al, 2004, p66). They suggest that, should the CO_2 fertilization effect be 'drowned out' by changes in climate, yield decreases of 9–22 per cent are possible under the SRES scenarios.

Response to CO_2 elevation may even be hard to detect against the background 'noise' of response to climate change itself, which may be nonlinear; for example, changes in rainfall patterns could alter the prevalence of pests and diseases, or demand changes in planting dates that might be constrained by photoperiod or temperature.

Alternatively, if it does make itself felt, it may do so in a negative way. For example, in 2007, *CO_2 Science* reported the results of Bhatt et al (2007), who had found that the dry-matter yield of the grass species *Cenchrus ciliaris* increased by 193 per cent with a CO_2 enrichment of 240ppm. 'Surely, such a response will be an *enormous* boon to the many people living in the arid and semi-arid tropics in the years and decades to come, as the air's CO_2 content continues its upward climb,' the piece says (*CO_2 Science*, 2007, p1). In contrast, Bhatt et al themselves had made no such statement, simply reporting their results – and with good reason; although *C. ciliaris* is an important fodder grass in some regions, in others (notably northern Australia and the south-west of the United States), it displaces native species and is regarded as a weed (see for example Tu, 2002, pp1–2). That such responses to CO_2 enrichment might not always be helpful has also been suggested by Morgan et al (2007), who found that elevated CO_2 levels would change plant composition in areas of steppe such as those found in Central Asia, with possible negative results for grazing. This occurred because the shrub *Artemisia frigida* was better able to use the extra CO_2 than existing grasses – a case where different photosynthetic pathways could make a difference.

Against this background, the consensus is that CO_2 fertilization either will not improve the food supply in those areas where that is needed, or that its effects under different moisture regimes and climate-change scenarios are highly uncertain. In fact, the idea that CO_2 fertilization will offset climate change or even make it beneficial is probably quite dangerous. As Canadell et al (2007, p375) put it, such errors could mean that: 'we are probably overestimating the potential future carbon sink of the terrestrial biosphere, and future climate change will proceed faster than currently predicted'.

The Earth Breathes Out

In fact, there is now broad agreement that 'feedback loops' such as CO_2 enrichment will eventually be positive – that is, they will add to climate change, not mitigate it.

By how much, is disputed. Berthelot et al (2005, p968) modelled terrestrial carbon changes over the next century under 14 GCMs and found reductions in the terrestrial carbon pool of between 30 and 240Gt; the differences are partially dependent on the behaviour of soil moisture in the tropics, with warmer and drier climates reducing NPP (op. cit., pp964, 966). In any case, as mentioned above, stimulation of NPP at higher temperatures will be outweighed by the faster decomposition of soil organic carbon (SOC) (Kirschbaum, 2000, p37).

A further variable is the behaviour of soil nitrogen (N); it has been argued that its faster mineralization in higher temperatures would increase NPP, but Houghton et al (1998, p26) argue that it is more likely to be immobilized in higher temperatures, as it will be used more quickly by soil micro-organisms – a process that will in itself consume soil organic matter. They quote an earlier study by Keeling that suggested a loss of 6Pg of terrestrial carbon per degree Celsius increase in temperature, equating to 3 parts per million of CO_2. 'As the Earth warms further, the positive feedback caused by temperature sensitivity could create a CO_2 source that equals or exceeds the current sink mechanism,' they state (op cit., p32). Loss of soil organic matter would play a key part in altering the balance. But because the rise in atmospheric carbon increases temperature more slowly than it does the CO_2 fertilization effect, there is a time-lag, and this may cause an apparent growth in the terrestrial sink for some decades (Scholes, 1999). The terrestrial sink will then 'flip' into a source; that is, instead of absorbing CO_2 from the atmosphere, it will start releasing more than it takes in.

How long this will take is a complex modelling problem, as it involves incorporating multiple carbon pools. Jones et al (2005) took the HadCM3LC GCM as their starting point and then coupled it with the well-established RothC model for soil carbon dynamics, instead of the single decay rate used initially (op. cit., p155). They anticipate that the terrestrial carbon pool will have released perhaps 10Gt by 2050 but about 70Gt by the end of the century (op. cit., pp163–164). They also point out that the CO_2 fertilization effect will eventually saturate.

Some terrestrial ecosystems could release CO_2 – or methane – very quickly; as discussed earlier, an example would be large peat bogs, especially those that had previously been frozen. This has led to the odd 'panic' story in the media. Even if these are discounted, however, it seems that the CO_2 fertilization effect is not going to mitigate either climate change itself, or its consequences for food supply, for long. Worse, the conversion of terrestrial systems from sinks to sources may start a process that cannot be reversed and may accelerate, with unpredictable consequences. David Viner of the Climatic Research Unit at the University of East Anglia has put it thus: 'When you start messing around with these natural systems, you can end up in situations where it's unstoppable. There are no brakes you can apply' (Sample, 2005 , p1). This also affects the argument put forward by commentators such as Bjørn Lomborg, author of the controversial 2001 book *The Skeptical Environmentalist*, that the optimal response to climate change is to accept it, and invest in adaptation and development. But we do not know what we are accepting, and it may be something to which many could not, in fact, adapt.

At the beginning of this chapter were two key questions that affect the validity of this book's hypothesis: whether climate change, induced in part by CO_2 emissions, was taking place; and whether agriculture could mitigate it. The answer to the first question is yes, but there is enormous uncertainty. The second question is dealt with in the chapter that follows.

Note

1 A distinction should be made between autotrophic respiration, in which the CO_2 returns to the atmosphere from the plant tissues itself, and heterotrophic respiration, which could involve the decomposition of organic matter which has already ceased to be part of a living plant. In the former case, the extent of respiration is directly significant to farm productivity; in the second it is

not in quite the same way, but does indirectly affect productivity, especially in the long term, and would be just as important in terms of climate feedback. As the IPCC points out (IPCC, 2001a, p190), the carbon all goes back into the atmosphere eventually.

2

Agriculture: Changing the Climate?

> *Empedocles taught that matter cannot originate from nothing and that matter cannot be destroyed. All happenings in the world are based on a change in form or on the combination or division of substances. Fire, air, water and earth are the four elements of which everything is composed. A continuous cycle is Nature's chief characteristic.* (Svante Arrhenius, Nobel lecture delivered on 11 December, 1903)

The question in the first chapter was whether climate change is happening and is worth mitigating, and the answer – despite scientific uncertainty – has been yes. The second question is whether agriculture can be used to mitigate it.

Carbon moves in a cycle between earth and atmosphere, changing its form but not its nature. CO_2 is created by the combination of carbon with oxygen through oxidation. The process of photosynthesis converts atmospheric CO_2 to sugars and carbohydrate, providing about 90–95 per cent of a plant's dry weight (Wittwer, 1995, p9); about 40–45 per cent of the plant material is organic carbon. Some of this will find its way underground as part of the plant's root system. When the plant dies, microbial activity mineralizes the residue back into CO_2, where it contributes to the atmospheric carbon pool and thus to climate change. So it makes sense to encourage its reconversion to plant material by increasing net primary productivity (NPP) – that is, biomass. Meanwhile the unexposed parts of the plant will be retained in the earth as soil organic carbon (SOC), which is contained in soil organic matter (SOM). The objective of 'carbon-friendly' farming is to ensure that there is as much of it as possible there at any one time. Doing this has benefits beyond carbon sequestration, as a high organic matter content can assist productivity and also, because it improves soil structure, water-holding capacity and resistance to erosion. However, SOM can take different forms in the same place, and will not all be cycled through a given patch of land at the same speed, so controlling – and measuring – this process is complex.

The first section of this chapter examines some issues related to organic matter in soils, drawing largely on Jenkinson (1988), Six and Jastrow (2002) and Six et al (2004). This is followed by a discussion of the extent to which agriculture can affect the size of the atmospheric carbon pool and whether there are practical implications for climate change. This is followed by a description of the main techniques advocated for increasing the size of the soil carbon pool.

Agriculture is defined as follows:

> *The word agriculture refers to a broad class of resource uses that includes all forms of land use for the production of biotic crops, whether animal or plants. In its broadest sense, agricultural land includes all land that provides direct benefits for mankind ... Only deserts; barren land; non-managed wetland, woodlands, and forests; and built-up areas are excluded.* (IPCC, 2000, p70)

In the context of the developing world, this means mainly cropland and grazing land. The latter are often rangelands, sometimes very extensive (for example, the great steppes of Central Asia); according to the FAO Statistical Yearbook for 2009, some 3.4 thousand million ha was pasture of one sort or another in 2007, roughly two and a half times the total area of arable land and a quarter of the total surface area.

Soil Organic Matter and its Dynamic Nature

If farmers are to be compensated in any way for sequestering or retaining soil carbon, or penalized for not doing so, then it is important to be able to predict just how much SOM will be accumulated, and at what annual rate, and for how long under a given agronomic regime. These questions will arise in the discussion of Kyoto and carbon trading mechanisms in later chapters. But they are worth some discussion here, as they relate to the dynamic and varied nature of organic matter in soils, which in turn underpins the C-friendly agronomic practices described later in this chapter.

The level of SOM in soil is a function of two main processes: the deposition of organic material below ground, and its subsequent mineralization.

The time between the two is the mean residence time (MRT). The longer this is, the more opportunity there will be to increase the SOM content. However, SOM does not last forever. 'In the long run, none [of the SOM created by photosynthesis] can withstand decomposition to carbon dioxide and water. If this were not so, any completely resistant fraction would by now cover the surface of the earth' (Jenkinson, 1988, p588).

The only exceptions would be anaerobic soils, in which microbial activity might be insufficient to break down some substances, in particular lignin (Jenkinson, 1988, p596). Thus substances such as straw might not decay in some conditions, in which case they would represent sequestered carbon for climate-change purposes but immobilized nutrients for the farmer. (In that example, use of the crop residues for biochar might be a better answer.) In any case, mineralization releases nutrients that are necessary for plant growth, especially in areas where there is little addition of mineral nutrients. So the objective should be to manage the process of mineralization, not stop it. As William Albrecht puts it in *Soils and Men*, the 1938 Yearbook of the US Department of Agriculture:

> *Attempting to hoard as much organic matter as possible in the soil, like a miser hoarding gold, is not the correct answer. Organic matter functions mainly as it is decayed and destroyed. Its value lies in its dynamic nature. ... The objective should be to have a steady supply of organic matter undergoing these processes for the benefit of the growing crop.* (Albrecht, 1938, p1)

For carbon sequestration purposes, however, the objective is not simply to have a 'steady supply', as Albrecht puts it, but to build up SOM more quickly than it is used by the crop until it reaches equilibrium – that is to say, the highest level that can be maintained consistent with the desired land use. The problem is that, while organic matter and carbon contents of a soil will clearly depend on the rate of input and the rate of output, the relationship between the two is unpredictable. For a start, the organic inputs are heterogeneous.

As Jenkinson (1988, p570) says: 'All the constituents of plants, animals and micro-organisms enter the soil at some stage, however transient their existence once there.' Estimating those inputs is therefore difficult. Even where they are known, carbon stocks may or may not display a linear relationship with them. Six et al (2000, p156) have pointed out that while some field experiments suggest soil carbon content proportional to inputs, others do not do so where carbon levels are high. They suggest a need for research into the mechanical explanations for saturation levels.

The mean residence time (MRT) – in effect, the speed with which organic matter turns over in soil – is also defined by complex factors that are imperfectly understood. Again, there is heterogeneity, with the different ages and types of organic materials. The most basic difference between types of organic matter in the soil is between those that have not yet decomposed, and those that have. At one end, one might find straw or newly dead animals or insects. At the other, one might find humus, which Jenkinson defines as 'fairly stable brown to black material bearing no trace of the anatomical structure from which it was derived' (Jenkinson, 1988, p564).

He describes (op. cit., pp604–605) a simple 'two-compartment' model in which the first compartment contains undecayed matter, the second the humus – or, to be more accurate, humic fractions; this is not completely homogeneous material. A given amount of fresh material enters the first compartment each year; this decays and feeds the second at a rate based on a marked deceleration of decomposition once about two-thirds of the material has decayed. When the SOM content reaches equilibrium – that is to say, when input and output are balanced – the average age of the organic carbon in the second compartment will be equal to the turnover time. 'In general, these simple exponential models give reasonably satisfactory representations of reality over the 10 to 100-year period, and have been successfully used to predict the effects of shifting cultivation on organic matter in tropical soils' (op. cit., p605).

However, this simple two-compartment model has drawbacks, not least the assumption that there are consistent rates of decomposition. Six and Jastrow (2002) point out that soils with similar average MRTs can have very different distributions of organic matter with different turnover rates, and that simulation models that account for these variations now generate 'more realistic descriptions of SOM dynamics' (Six and Jastrow, 2002, pp939–940). Thus crop residues with a high lignin content might decompose relatively slowly, while root material from legumes might become part of the humic fraction quite quickly. In either case, the extent and type of soil microbiota play a role, as does the soil type. After decomposition, MRT may also be highly variable depending on the soil structure. The extent to which the SOM is exposed to soil biota can depend on the extent of aggregation, with that in microaggregates markedly better protected than in macroaggregates. The difference may be greater in some farming systems than in others, but can be significant. Six and Jastrow (2002, p940) quote studies that indicate an MRT of 60 years in tropical pastures for organic matter associated with aggregates of >200µm and 75 years for those with <200µm, but in a wheat-fallow system the range could be from 8 years (250–2000µm) to 79 years (53–250µm); with corn (maize), 42 years (>250µm) up to 691 years (50–250µm).

This might imply that the presence of microaggregates can assist in the estimation of MRT, and that macroaggregates are of less importance. In fact, microaggregates containing SOM will form *within* macroaggregates that contain plant material (Six et al, 2004, p12). Moreover the presence of macroaggregates implies pore size that will assist both biotic activity and water ingress – both needed for a healthy crop, which will in turn provide the organic matter. When assessing the probable MRT of SOM in a given farming system, therefore, it may make more sense to look at the extent to which it encourages aggregation in general rather than paying too much attention to microaggregates. It is therefore logical that no-till systems, in which macroaggregates are not broken up by the plough, increase MRTs, and indeed there is evidence of this (Six and Jastrow, 2002, pp939–940; Six et al, 2004, p12).

This is still not the whole story. Even where macroaggregates are vulnerable to disturbance, organic matter held in microaggregates will still have a longer MRT. A number of factors affect the formation of such aggregates, and estimates of MRT based solely on the crop and agronomic practices will not necessarily be accurate. Six et al (2004, p8) identify soil fauna, soil micro-organisms, roots, inorganic binding agents (oxides and calcium), and environmental variables (such as freezing and thawing, wetting and drying, and fire). Each of these has variables of its own. For example, the presence of earthworms can greatly increase the formation of microaggregates containing plant-derived carbon – but this will not automatically be so; it depends on the type of soil and earthworm species, and where there is a positive long-term effect on carbon stocks, it may be preceded by a decline (Six et al, 2004, pp17–18).

None of the above means that carbon sequestration and mineralization *cannot* be modelled; in fact it will have to be for baseline-setting and carbon payments, areas discussed in Chapter 5. But the agronomic practices described later in this chapter are recommended because they increase organic matter input (intensification) or because they increase MRT by reducing exposure of SOM to air and disturbance, often by preventing destruction of soil aggregates (no-till, erosion control). It is therefore important to stress the uncertainty (and site-specificity) that may surround their exact potential for carbon sequestration or retention, as well as the global estimates for carbon sequestration potential in agriculture discussed below. In the meantime, research continues into the behaviour of organic matter and carbon in soils, and the accuracy of these projections is likely to improve.

Soil Carbon: How Much, Where, and *Potentially* How Much?

According to the IPCC (2001a, p192), global terrestrial carbon stocks amount to between 2221Pg C and 2477Pg C, depending on which estimate is used. Of this, 657Pg C is held in plants and 1567 as soil carbon (ibid.). These are substantial amounts when compared to the atmospheric carbon pool, which is about 770Pg C (Lal, 2002, p353).

It is the soil carbon that is most important for this book, as it constitutes the carbon sink for which farmers might be paid. However, carbon held as plant material is also vital; although it may only exist for part of the year, it is mitigating climate change for that period – indeed there are large seasonal variations in atmospheric CO_2. In any case, some plant material is semi-permanent, as in the case of trees; even when cut down, these may remain part of the terrestrial carbon pool as wood products. So it is not possible to ignore it. Figure 2.1 shows the terrestrial carbon stocks broken down three ways.

Further variation arises from the fact that although a forest contains a lot of above-ground carbon, turnover can be slow, so it does not recreate a large amount of it year on year. As described earlier, the biomass that *is* recreated in a given biome is its net primary production, or NPP.

A biome with high NPP is likely to present more opportunities for the sequestration of carbon, for example by using plant species that leave more of their mass underground as root systems, or by practising soil conservation that will reduce disturbance and exposure of soil carbon. On this basis, the two areas of interest to this book – tropical savannah/grassland, and cropland combined – offer as much potential as tropical

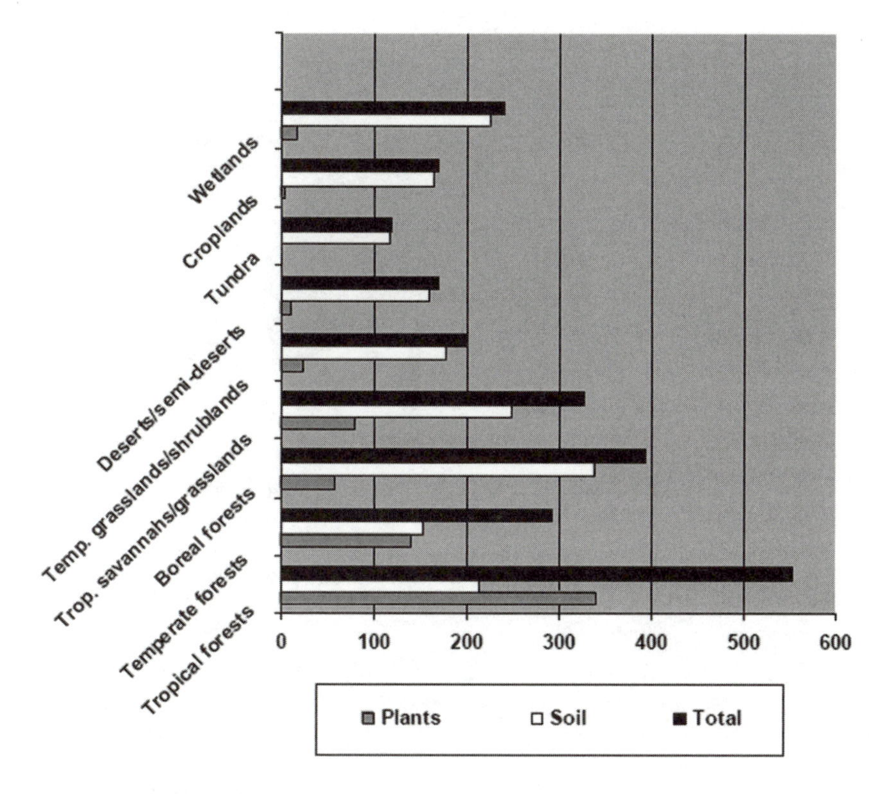

Figure 2.1 *Global terrestrial carbon stocks (Pg C)*

Source: IPCC (2001a, p192)*

* The IPCC quotes two sets of data for terrestrial carbon stocks. One is from the German Advisory Council on Global Change (WGBU). The second takes plant stocks from Mooney et al (2001, quoted in IPCC, 2001a, p192) and soil from the Data Information Service of the International Geosphere-Biosphere Programme. For simplicity, only the second, more recent set is used here, but that does not include the figure for wetlands, which is taken from the WGBU.

forests. At the same time, these types of land use/biome also store a far higher percentage of their carbon stocks as soil carbon, a more permanent type of sink. This can be seen clearly in Figure 2.1.

Although these biomes would not hold more carbon per ha than forests, they can *accumulate* it more quickly, as the carbon cycle is much slower in forests. (This can be seen in Figure 2.2, which shows the distribution of NPP – that is, production of plant material and, therefore, sequestration of carbon (as opposed to its retention). Moreover, grassland and cropland,

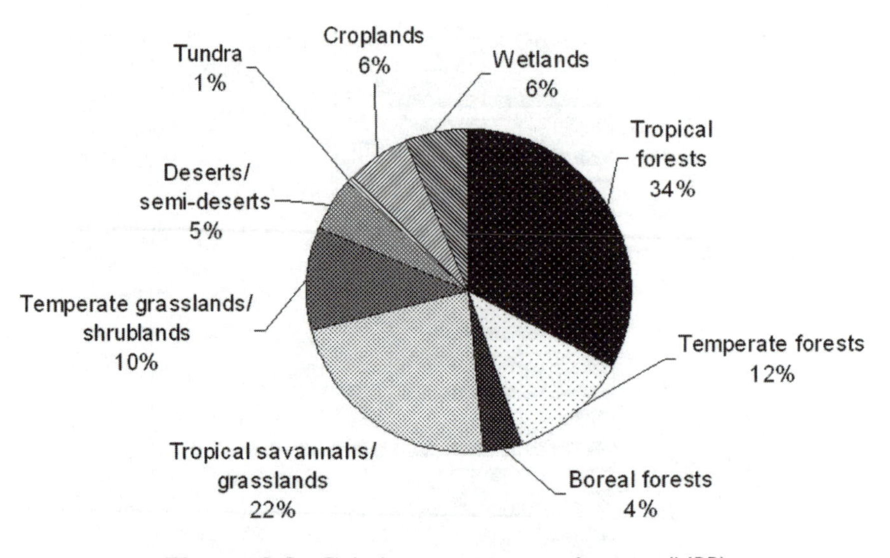

Figure 2.2 *Global net primary productivity (NPP)*

Source: IPCC (2001a, p192)*

* The IPCC quotes two sets of data for NPP; from Atjay et al (1979, quoted in IPCC, 2001, p192) and from Mooney et al (2001); the second, more recent, set has been used here, but that does not include the figure for wetlands, which is taken from Atjay et al.

unlike tropical forest, is producing food on a large scale, whereas forests – unless commercially logged – only directly support those who live in them (although they do provide crucial ecosystem services such as water quality – but agriculture has a role in that too). Sequestration in grassland and cropland is therefore better able to support broader development objectives besides climate-change mitigation.

How much could agriculture add to the terrestrial pool? Soil can lose carbon as well as accumulate it and that, historically, is what it has done, adding to rather than mitigating climate change. This is because a piece of land that is producing food is burning carbon to meet human needs in broadly the same way as a factory run off fossil fuel, and the amounts of energy involved can be comparable. Once again, this is well expressed by Albrecht:

> *An acre of the better Corn Belt soil in Iowa ... burns carbon at a rate equivalent to 1.6 pounds of a good grade of soft coal per hour ... A 40-acre cornfield during the warmer portion of a July day is burning*

> *organic matter in the soil with an energy output equivalent to that of a 40-horsepower steam engine. In other words, every acre (a little under half a hectare) is effectively a factory using the equivalent of one horsepower.* (Albrecht, 1938, p1)

Nearly 70 years on, this is still an elegant visualization of the role soil has in climate change. Yet, by creating plant material through photosynthesis, the process is also *removing* carbon from the atmosphere and converting it into fuel at the same time, something that a factory does not do. Regulate this process for optimal soil carbon content, and soil carbon will be built up rather than expended. The problem is that, up to now, the balance in agriculture has been wrong: it has been a net user of carbon, through the destruction of forests for farmland and the subsequent mineralization of soil C.

According to the IPCC, land-use change in general is a significant source and potential sink. 'Hypothetically, if all of the carbon released by historical land-use changes could be restored to the terrestrial biosphere over the course of the century (e.g., by reforestation), CO_2 concentration would be reduced by 40 to 70 ppm,' it states (IPCC, 2001a, p12).

Agriculture should certainly be part of this. Exposure of organic matter through ploughing or erosion causes decomposition, so that much carbon is converted into CO_2 by respiring micro-organisms. Intensive agricultural use, without appropriate management, can cause SOC content to fall further. Lal and Bruce (1999, p178) put the total historic loss from agricultural soils at 55Pg C.

This may not sound much against the total global soil organic carbon pool, which as stated above is about 1567Pg C, or even the atmospheric pool at 770Pg C. However, what matters is not the percentage of total soil or atmospheric carbon, but the percentage of the anthropogenic *increase* in CO_2 that can be offset. Total atmospheric carbon increased from about 613 to 780Gt between 1850 and the early 20th century (Carbon Dioxide Information Analysis Center, 2003). If, as Lal and Bruce (1999, p178) estimate, about 75 per cent of the lost 55Pg C could be recovered, this is a worthwhile proportion of that anthropogenic increase. It looks the more so if measured against ongoing emissions. The net annual increase in atmospheric CO_2 during the 1990s was 3.2 ± 0.1Pg/yr C (IPCC, 2001a, p39). Lal and Bruce (1999, p182) think that the carbon sequestration potential of world cropland is about 0.43–0.57Pg/yr C, or very roughly 20 per cent of the annual increase.

The IPCC working group on agricultural mitigation for the Fourth Assessment Report estimates that agriculture is responsible for 5.1–6.1Gt carbon dioxide equivalent (CO_2e) annually, or about 10–12 per cent of anthropogenic GHG emissions (IPCC, 2007a, p499). However, the majority of this is, they say, N_2O – and CH_4 – nitrous oxide and methane, not carbon. They state that although the total carbon exchange between the atmosphere and terrestrial carbon stock is very large, net emissions are quite small – about 0.04Gt (ibid.). But it follows that, with very large fluxes, there may be mitigation opportunities, and the working group report estimates that 89 per cent of the technical mitigation potential comes from carbon sinks. It estimates that potential at approximately 5500–6000Mt/yr CO_2e up to 2030, but puts economic potential – that is to say, the amount that might be sequestered in practice – at anything from 1500 to 4300Mt/yr, depending on the carbon price (ibid.). As always with the IPCC, a qualifier is attached to these estimates; the technical potential, for example, is regarded as 'medium agreement, medium evidence'.

But the figures above refer to the global terrestrial stock and make no distinction between rich countries and poor. Tropical forests, to be sure, are largely in developing countries, as is much of the world's grassland and rangeland; but in the case of cropland the geographical distribution of the carbon pool is less clear. There are too few data on this, and comparing different sources is problematic, as the assumptions are almost never the same, not least on how much land is likely to be managed in a carbon-friendly manner. However, a few estimates do exist and can provide some indication.

The working group report itself says that about 70 per cent of the potential is in non-OECD/EIT[1] countries (ibid.). Although it gives regional breakdowns for total mitigation potential in agriculture, its estimates for non-livestock options are presented by climatic zone instead, and do not divide neatly into developed/developing regions. The estimates are for potential by farming practice; although the IPCC gives a low, mean and high range, only the means are given here. They are presented in Table 2.1. The IPCC also includes options that constitute fundamental land-use change, such as set-aside and agroforestry, but these are omitted here; they are important, but should be seen as a separate subject, with the focus in this book being sequestration activities within existing farming systems. Also, the working group report pays more attention to climate mitigation through reduction of other greenhouse gases, such as N_2O and CH_4 from

Table 2.1 *Potential for agricultural carbon sequestration over 20 years, by climate zone (t/ha/yr CO_2)*

Practice	*Cool-dry*	*Cool-moist*	*Warm-dry*	*Warm-moist*
Agronomy	0.29	0.88	0.29	0.88
Nutrient management	0.26	0.55	0.26	0.55
Tillage/residue mgt.	0.15	0.51	0.33	0.70
Water management	1.14	1.14	1.14	1.14
Restoring degraded land	3.45	3.45	3.45	3.45

Source: adapted from IPCC (2007a, p512)

fertilizer and livestock. These emissions will be discussed briefly later in this chapter; they are a separate subject from carbon sinks. (That said, CO_2 sequestration should clearly not be by methods that would increase emissions of other greenhouse gases, at least not to the extent that there is no net mitigation. This is a problem that will recur in this book.)

However, as stated above, these figures do not distinguish between the developed and developing world. For the latter, there are fewer estimates. Niles et al (2003, pp78–79) estimated the annual carbon-sequestration potential from sustainable agriculture in 2003–2012 for 14 countries in Latin America, 23 in Africa and 11 in South Asia. The methods assumed for sequestering carbon included no-till, cover crops, composts and others, and these are discussed later.

The aggregate totals (Niles et al broke them down by country) are given in Table 2.2.[2] The end of this period is now quite close, but this work is being quoted to indicate how much carbon might be sequestered in developing countries as opposed to richer ones. In percentage terms it is

Table 2.2 *Potential for agricultural carbon sequestration in the developing world, 2003–2013*

Region	*Adoption (Mha/yr)*	*Mt/yr C*	*Value in US$m*
Latin America	11	93.1	644.1
Africa	15.9	69.7	482.1
Asia	18.5	227.3	1572.9
Total	49.6	420.6	2910

Source: adapted from Niles et al (2003, p79)

much less, but still amounts to nearly half a million tonnes in less than 10 years. Moreover, its cash value is proportionately more important to those that would sell it. Most important, however, these are the low-latitude, hot areas that, with higher temperatures, might 'flip' from carbon sources to sinks, as described earlier. If farming protects these especially fragile sinks, the global benefits from reduced emissions through sustainable practices may greatly exceed those now apparent.

Several points need to be made with regard to Niles et al's figures. First, they are based on a 'medium' sequestration rate; as the authors point out, the potential for given agricultural land uses can be lower or higher, depending on the level of practices used – the authors give estimates (Table 2.3). Second, the projections are predicated on notional adoption rates; those the authors quote seem reasonable, but they are hard to predict and small variations could greatly increase the carbon sequestered. Moreover, adoption rates could themselves be distorted by any carbon funding that might be offered.

Third, the figures are dominated by certain large countries; out of the potential 420.6 million tonnes for Asia, China's potential is estimated at 84.3 million tonnes, for example, while for Bangladesh, despite its large size and population, it is 5.3. Laos has just 0.6. In Latin America, Mexico and Brazil dwarf the others. The balance is more equal in Africa. Even so, Brazil, India and China may be expected to take the lion's share – not least because they will have the infrastructure to mount the projects, so their share of the agricultural sequestration budget could be even higher than these figures indicate. The funding available for countries such as Mali and Laos may be limited. These issues are discussed in a later chapter.

Finally, the study assumes that only 10 per cent of the world's grasslands can be improved. This may be the case; however, the area of grassland

Table 2.3 *Carbon sequestration potential* (t/ha/yr C) in four agricultural systems under different levels of management*

System	*Low*	*Medium*	*High*	*Very high*
Arable	0.3	0.65	1.3	3.1
Rice paddy	0.1	0.1	0.1	0.1
Permanent crops/agroforestry	0.4	0.5	0.7	0.9

* Includes above-ground biomass
Source: adapted from Niles et al (2003, p76)

and savannah in the developing world is so enormous that, again, any error here would involve huge amounts of carbon – and 71 per cent of the world's grasslands are thought to be degraded to some extent (IPCC, 2000, p205).

It is also noticeable that Niles et al have excluded the Arab countries (but not Sudan), Iran, Pakistan and the former Soviet states of Central Asia; the latter, although sometimes considered as transitional countries, did not accede to the Kyoto Protocol as Annex 1 (i.e. developed) countries, and should probably be seen as developing for our purposes. The excluded area contains more than 500 million ha of semi-arid steppe and rangeland, an area whose carbon budget could dwarf even that of China's grasslands. The IPCC (2000, p199) suggests that a combination of grazing management, species introduction and fire management of dry tropical grassland could accumulate 0.9t/ha/yr C; the IPCC assign this estimate 'low confidence', but if 10 per cent, or 50 million ha, accumulated 0.9t/ha for 40 years (the period the IPCC thinks it would take to reach saturation point), this would be 1.8Pg C. At the carbon price used by Niles et al of US$10/t, this would be worth US$18,000 million – which, even over so long a period, might certainly be enough to finance appropriate restoration and conservation activities.

None of this undermines the value of Niles et al's exercise, which provides helpful indicative data on the potential of sinks in developing countries – something that had hitherto been neglected. Besides, the authors themselves say that further research is needed (Niles et al, 2003, p85).

Before going on to consider how much carbon might be sequestered by agricultural land, it is important to note that agriculture can mitigate climate change not just by offsetting emissions from other sources, but by reducing its own. Land-use change, mainly the destruction of forests for agricultural production, has been a significant source of emissions since settled agriculture began; but there are others. According to the IPCC working group, CH_4 and N_2O emissions from agriculture grew by nearly 17 per cent between 1990 and 2005 (IPPC, 2007a, p499), although estimates vary.

Smith et al (2007, p14) report an overall growth of GHG emissions from agriculture of 14 per cent between 1990 and 2005. The two fastest growth areas were N_2O from soils at 21 per cent and N_2O from manure management at 18 per cent; CH_4 emitted by ruminants from enteric fermentation increased nearly as fast at 12 per cent, but in absolute terms N_2O increased almost twice as rapidly at 31Mt of CO_2e a year (ibid.).[3]

This growth has been driven by the developing, not the developed, world. In 2005, those regions consisting mostly of non-Annex 1 countries (that is, mainly developing countries that have no emissions caps under the Kyoto Protocol) accounted for 74 per cent of global agricultural emissions (op. cit., pp12–14). This will continue; OECD countries are expected to contribute only 23 per cent of N_2O emissions by 2020, down from 32 per cent in 1990 (US-EPA, 2006, 5-3 to 5-4). At the same time, N_2O emissions from China, from other centrally-planned economies of Asia,[4] and South and East Asia are expected to grow by more than 50 per cent, and those from Africa, Latin America and the Middle East by over 100 per cent. 'These regional increases are driven largely by projected emissions increases in China, Brazil, Argentina, Nigeria, Bangladesh, India, and Iran,' says the US-EPA (ibid.). 'Only a handful of OECD countries are expected to show increased emissions [by] 2020.'

These figures raise two important points. The first is that, as far as agriculture is concerned, the industrializing countries of the developing world do bear some practical responsibility for GHG emissions and their mitigation. To what extent they have a *moral* responsibility is a different question. (These arguments are discussed briefly in a later chapter.) But it is clear that, if mitigation activities in agriculture are to have any impact on climate change, they must take place in the industrializing non-OECD countries at least as much as in the OECD.

The second point is that N_2O emissions from soils and manure management are an important driver of climate change. Activities to sequester CO_2 could, if badly planned, increase them (for example by attempting to raise NPP using N fertilizer). As Lal (2009, p164) states, any estimate of net carbon sequestration must take into account the fluxes of all GHGs. But with careful planning, the extra N_2O emissions can be offset by the extra carbon. Better still, well-designed activities could sequester CO_2 and *reduce* N_2O emissions at the same time, not least by reducing erosion that has an impact on soil fertility, increasing the need for nitrogen fertilizer. However, the field survey described in the later chapters will look at the practicalities of this. For now, the rest of this chapter will describe the management practices that would be expected to increase the carbon sink in agriculture.

How Carbon Can be Sequestered in Agriculture

As stated earlier, much plant material is incorporated into the soil as soil organic matter (SOM), over half of which is soil organic carbon (SOC) (Follett, 2001, p88; Post et al, 2001, p73). (The US Department of Agriculture suggests SOC × 1.72 = SOM as a yardstick, although this is not universally accepted.) Anything that increases biomass, therefore, will result in seasonal reductions in atmospheric CO_2, but the real challenge is to get organic carbon underground and keep it there, where it will remain in the terrestrial pool all the year round. As Batjes (2001) puts it:

> *Recommended management practices to build up carbon stocks in the soil are basically those that increase the input of organic matter to the soil and/or decrease the rates of soil organic matter decomposition. These practices will generally include a combination of the following: tillage methods and residue/stubble management; soil fertility and nutrient management; erosion control; water management; and crop selection and rotation.* (Batjes, 2001, p136)

The IPCC (2000, p200) quotes studies that suggest that most agricultural soils have not reached saturation point and are potential sinks. It divides the strategies for sequestration into three broad groups, as described below.

Agricultural intensification

This is anything that increases biomass production, including rotations, nutrients, improved cultivars and irrigation (IPCC, 2000, p202).

Conservation tillage (CT)

That is to say, reducing or eliminating ploughing. The definition used by the IPCC is that of the Conservation Tillage Information Center (CTIC) and calls for crop residues to remain on the soil surface after planting; the amount or percentage of residues that must be used in this way depends on whether water or wind erosion is seen as the primary threat (IPCC, 2000, pp202–204). Certain forms of tillage are acceptable if they are compatible with these requirements for crop residues. But any reduction in inappropriate tillage should reduce organic matter breakdown, and thus slow loss of soil carbon, even if it does not increase it. Some caution is needed here;

as Giller et al (2009) have pointed out, there are some problems with the tillage regimes advocated for carbon accumulation, especially where they are transferred to farmers as rigid packages of technology. This will be discussed in Chapter 8, when the 'carbon-friendly' practices are reviewed in the context of the field survey.

Erosion control

The IPCC quotes Lal and Bruce's estimate that erosion displaces about 0.5Gt of soil carbon per year, of which about a fifth enters the atmospheric CO_2 pool (Lal and Bruce, 1999, p179). It calls for measures such as terracing and shelterbelts (IPCC, 2000, 4.4.2.3, Fact Sheet 4.4). Again, there is a need for caution here. The question of the relationship between erosion and the carbon cycle is discussed later in this chapter.

These definitions are a little broad, but narrower definitions may be found in the literature. Follett (2001) mentions several strategies; they are effectively covered by the IPCC, but he is more specific, citing increase of land cover through winter cover crops, nutrient inputs and supplemental irrigation.

Lal and Bruce (1999, p182) estimate the potential gains for strategies that fit broadly under the IPCC umbrella (see Table 2.4), plus soil restoration – the IPCC says little about the potential for rehabilitating degraded cropland, but carbon funding might make it economic.

There are two interesting omissions from the table. One is intensification, as nobody really knows how much carbon could be sequestered in this way, or how much would be lost through leakage. There is also a complex relationship between crop rotations and tillage regimes. These factors are discussed in Chapter 8. The other omission (because it has only

Table 2.4 *Carbon sequestration potential of strategies for arable land (Pg/yr C)*

Soil erosion control	0.08–0.12
Soil restoration	0.02–0.03
Conservation tillage and residue management	0.150–0.175
Improved farming/cropping systems	0.18–0.24
Total	0.43–0.57

Source: Lal and Bruce (1999, p182)

recently been discussed in this context) is biochar. This is a potentially very important new field in carbon sequestration, but at the moment raises as many questions as it answers. It is discussed at the end of this chapter along with another, much older subject that is included in the table, but is also rife with uncertainty – erosion control.

A key point about all these strategies is that they should benefit farmers anyway, even without carbon-related financial support. Erosion control clearly has implications for sustainable productivity. So does grassland management. But increasing SOM and therefore soil carbon is – with one or two reservations – good for crop production in itself. Lal and Bruce make this point strongly, arguing that 'increase in SOC content improves aggregation, plant available water capacity, ion exchange capacities, soil biodiversity, and soil quality' (Lal and Bruce, 1999, p178). Authors *not* considering climate change have also noted the benefits of increasing SOM (and effectively SOC). McDonagh et al (2001, p13) point out that: 'Levels of SOM have been shown to be particularly important in cultivated tropical soils where the high temperatures lead to rapid SOM breakdown, organic matter reserves are often low and the use of other inputs rare.' Altieri (2002, p6) comments that 'soils with high organic matter and active soil biological activity generally exhibit good soil fertility as well as complex food webs and beneficial organisms that prevent infection'. Organic matter performs a particularly important function in assisting the aggregation of soil, improving the soil structure, increasing water infiltration and reducing erodibility. The importance of aggregation has already been described.

It is worth repeating Albrecht's stricture, quoted earlier, that simply hoarding organic matter is – from a production point of view – pointless. SOM as a fuel for plants is no different from petrol in a car's tank – it produces nothing until it is burned; as Arrhenius said, everything that happens, does so when something changes form, divides or combines with something else. It is useless building up organic matter in the form of immobilized nutrients, for example crop residues that do not decompose – which they may not do in less humid climates. In those situations, it might make more sense to use crop residues as feed, biomass for energy – or as feedstock for biochar (discussed below).

With this reservation, however, high levels of SOM – and, therefore, of SOC – are good for agriculture. In fact Lal (2004a, p1626) reports findings that a 1t/ha increase in SOC has led to an extra wheat grain yield of 27kg/ha in the Great Plains of the United States, 40kg/ha in the semi-arid pampas of Argentina, and 6kg/ha of wheat and 3kg/ha of maize in alluvial soils of northern India. Maize yield increases were 17kg/ha in Thailand and 10kg/ha of maize and 1kg/ha of cowpea in Western Nigeria. Such increases are obviously subject to many variables, and are perhaps small compared with those that might be obtained with chemical inputs. But they are gains and not losses and, besides, they should be seen as part of an overall improvement to sustainability. If carbon funding can secure such gains, there is in theory a clear win-win situation.

The Erosion Question

This is also true of controlling soil erosion. It does have an impact on productivity. Lal (2001) has estimated production losses for Africa in 1989 at 8.2 million Mg of cereals and 9.2 million Mg of roots and tubers; he also argues that global crop-yield losses due to erosion could be '10 per cent in cereals, 5 per cent in soybeans, 5 per cent in pulses, and 12 per cent in roots and tubers' (Lal, 2001, p531). At the same time, the amounts of soil translocated can be high, suggesting that this loss of food production is accompanied by carbon emissions.

However, erosion control presents huge challenges, and has a difficult history in agricultural development. Some of the practical problems are discussed in a later chapter, but for now it is worth briefly discussing the theoretical uncertainty surrounding its relationship to climate change. How much carbon is really emitted this way?

One would need first to know how much soil is being moved around, and this is a question in itself. As Stocking (1996, p327) says, when environmental policy or development actions are in the frame, 'erosion assessments should themselves be audited as to who has made them, how, and why'. An example of this is the publication in 1984 by the Worldwatch Institute of an estimate that global annual soil erosion was 26 billion tonnes above the level that would not threaten agricultural production. Gisladottir and Stocking (2005, p102) comment that: 'This is a huge mass of soil, which, if true, would remodel whole landscapes within historical record.' They report that this example was caused by conjecture based on too little data, and go on to explain why some estimates of worldwide soil erosion are inordinately high; these include a tendency to scale-up runoff data from small plots, a process which ignores the scale-dependent nature of such measurements and gives gross overestimates if extrapolated to a regional or watershed basis (ibid.).

When global figures are quoted, this gets worse. 'Figures are routinely extrapolated, often by scientists, to give a global overview of soil loss. They persist even today for the continuing debate on land degradation and desertification,' they state (op. cit., p103), and are critical of the hyperbole that surrounds such estimates. They quote as a further example the statement by a former director-general of one of the international agricultural research centres that: 'The livelihoods of more than 900 million people in some 100 countries are now directly and adversely affected by land degradation' (El-Beltagy, 2000, quoted in Gisladottir and Stocking, 2005, p103).

Both of the examples in the paragraph above came from specific contexts that might demand an emphasis of the dangers of soil erosion. The first was from the Worldwatch Institute, a Washington-based environmental think tank that says on its website that its 'work revolves around the transition to an environmentally sustainable and socially just society—and how to achieve it'. That does not of course mean that either of the statements quoted above were made in bad faith; there is no reason to think that. But there is always a temptation to pick data that backs one's cause.

Apart from anything else, soil erosion is hard to estimate on a global scale simply because it is so site-specific. Pimentel (2006, p123) reports that erosion on cropland averages about 30t/ha/yr, but locally it can be lower – or much higher: 'In the Philippines, where more than 58 per cent of the land has a slope of greater than 11 per cent, and in Jamaica, where 52 per cent of the land has a slope greater than 20 per cent, soil erosion rates as high as 400 t/ha-yr have been reported' (Pimentel, 2006, p120).

However, some rough global estimate may be necessary to enable its prioritization or otherwise in the climate-change agenda. Not long after the IPCC figures quoted earlier were published, Lal (2003, p437) suggested that the amount of soil displaced is far greater and that the total emitted as CO_2 could be as high as 0.8–1.2Pg/yr C. If this is correct, soil erosion emits the equivalent of about a third of the annual increase in CO_2, and erosion control is an even more important strategy than the IPCC suggests. If one does accept Lal's contention that as much as 20 per cent of translocated SOC is mineralized, this could translate into as much as 50t/ha carbon. Multiplied by 3.66, this is 183 tonnes of CO_2e. If it is sold as a straightforward certified emissions reduction (CER) to a secondary buyer and fetches up to $US7 a tonne – which is possible – this is a startling US$1281/ha/yr.

No one is really going to recoup this sort of sum. There would be monitoring and transaction costs, and investment up-front would be needed to produce saleable CERs from such a project – an area discussed in Chapter 5. Besides, the 20 per cent figure is an average, and every site would presumably be different in terms of how much carbon ended up in depositional sites or aquatic ecosystems (in which it may not be mineralized).

Moreover, although the IPCC has in the past mentioned the potential for erosion control to reduce emissions, the working group on agricultural mitigation for the fourth, most recent, assessment report (IPCC, 2007a) does not say much about it. This may be because uncertainty as to soil erosion's role seems if anything to be growing. Berhe et al (2007, p338) point out that while some studies have concluded that soil erosion is a net source of emissions, others have decided that it is a net *sink*. They quote Stallard's (1998) calculation that soil erosion leads to a human-induced sink of 1.5Pg/yr C. They themselves conclude that erosion-induced carbon sequestration would have offset 10 per cent of fossil-fuel emissions in 2005. Van Oost et al (2007, p628) also conclude that erosion from agricultural lands is a net sink, although in their view it is so small as to be marginal.

Why the disagreement? One reason is that as SOC is transported away from a site by erosion, NPP at the source replaces it, and there can be different assumptions about the speed with which this happens. Lal and Pimentel (2008), in a response to Van Oost et al, argue that NPP at the erosion source is likely to be reduced by the degradation that has taken place. This is clearly logical, but Van Oost et al (2008, p1042) respond that they have accounted for this. Moreover, Berhe et al (2007, p341) point out that, whereas organic carbon in agricultural soils eventually reaches saturation point and can sequester no further CO_2, the loss of carbon from erosion brings it back below saturation point so that the process may begin again – so that, if the original, translocated, carbon has not been emitted but is protected in a depositional site or aquatic system, there will be a net sequestration.

There will be variables that affect this. One is whether the original carbon has indeed been protected elsewhere and not mineralized during transport; we do not know, on a global scale, to what extent that is the case. Another is that NPP at the eroded site may depend on the inputs a farmer would want to expend on what may be a sloping and vulnerable location. The fact that land can be improved does not mean it is in the farmer's interest to do so – a subject that will be important later in this book.

Recently, Kuhn et al (2009) have pointed to other factors that should have been considered. They argue that more needs to be known about the fate of translocated carbon as it moves through the landscape – how much is mineralized, and how its composition is affected by the agricultural land through which it passes. Previous research, they state, has been based on the assumption that the soil carbon content is the same when it reaches a depositional site as it was when it was at its point of origin, regardless of what has happened to it on the way. Second, they say that inter-rill erosion, in which the exposed SOC may not go far and is mineralized on the soil surface, may have an important role. Last but not least, they argue that the method of estimating erosion through the movement of ^{137}Cs, as used in studies such as Van Oost et al's, yields a 50-year mean but does not take into account the fact that erosion is a non-linear process that will decline as the most erodible soil gets moved, and that this has been happening since settled agriculture began; the same therefore applies to the soil carbon within it. Arguably, this question – the validity of an observation at point(s) in time for measuring ongoing and possibly non-linear processes – is important for the study of climate change in general.

In any case, while a global figure might be fixed, the local percentage would surely be site-specific depending on the presence of watercourses and depositional sites. It may be that the whole question is moot. Even if researchers eventually establish whether anthropogenic soil erosion is a source or a sink, this will be of only so much use for climate mitigation; what is important is the net flux *within a given watershed.* After all, no mitigation intervention will be undertaken consistently across the globe in every country from Canada to North Korea.

In 2005 the author asked Professor Lal what he now thought was the percentage of translocated SOC emitted as CO_2. 'I don't know,' he replied, 'but I am trying to find out.' It is an answer that could have been given for almost any aspect of climate-change research. To be sure, the debate between soil scientists has been of a much higher order than that between climate-change 'sceptics' and their critics. The latter is a bad-tempered, politicized, dishonest and sinister debate that will resolve nothing. The debate between soil scientists, by contrast, is a courteous exchange between professionals who genuinely seek answers. Yet they both raise the question of scientific uncertainty, a topic that is important to the monitoring and verification issues discussed in Chapter 6.

Biochar: An Exciting New Opportunity?

Biochar is a more recent arrival in climate-change discourse. But it too involves much uncertainty, although this is largely because there has not yet been time for much research in the climate-mitigation context. It concerns pyrolysis – that is to say, the heating of biomass in an environment that is free or nearly free of oxygen. This is how charcoal is made. When the biomass is heated in this way in order to be used on the soil, it is called biochar (Lehmann and Joseph, 2009, p1). The feedstock will be whatever is available.

The rationale for biochar in a climate-change context is clearly presented by Lehmann et al (2010, p344):

> *The main incentive of biochar systems for mitigation of climate change is to increase the stability of organic matter or biomass. This stability is achieved by the conversion of fresh organic materials, which mineralize comparatively quickly, into biochar, which mineralizes much more slowly. The difference between the mineralization of uncharred and charred material results in a greater amount of carbon storage in soils and a lower amount of carbon dioxide, the major greenhouse gas, in the atmosphere.*

Biochar is, in principle, not new. Justus Liebig, the great German researcher who is associated with (amongst other things) the role of nitrogen in agriculture, observed such a technique in 19th-century China, and biochar has also been used to improve golf courses (Lehmann and Joseph, 2009, p4). However, interest in this area has been stimulated in recent years following the work of a Dutch researcher, the late Wim Sombroek. Sombroek, with others, described the existence of large areas of very rich soil in the Amazonian basin, with a very high organic matter content and good cation exchange capacity. Well known locally, these soils are referred to as *Terra preta de indio*, although Sombroek distinguishes between *Terra preta* and *Terra mulata*, a slightly less rich soil that often surrounds areas of *Terra preta* (Sombroek et al, 2002, p1935-2). They appear to have been created by a combination of human habitation and deliberate land-use practice, and are also referred to as anthropogenic black earth. 'The data available thus far point to a much higher soil organic matter (SOM) content of the two soil types – often double the normal amount – which has moreover a high degree of permanency' (op. cit., p1935-4). Sombroek et al state that it is not clear how these soils were created, and suggest that, in view of their productivity and high carbon-storage, it would be useful to try and replicate them.

There appears now to be great enthusiasm for this. Soil qualities associated with biochar use include improved soil structure and aeration (Kolb, 2007, quoted in Downie et al, 2009, p22); soil aggregation and its associated carbon storage can be encouraged by biochar's biological properties, including the action of mycorrhizal fungi and other soil biota (Thies and Rillig, 2009, pp93, 101). Biochar as a soil amendment has been found to reduce leaching of N, saving fertilizer (and thus presumably also N emissions), and to increase crop productivity by 38–45 per cent – much more on poor soils such as those on which Lehmann and Rondon experimented in Colombia (Lehman et al, 2003 and Lehmann and Rondon, 2006, quoted in McCarl et al, 2009, p349). But the real driver behind the current interest in biochar is its climate-mitigation potential.

Woolf et al (2010), in a recent paper in *Nature Communications*, suggested that: 'sustainable global implementation of biochar can potentially offset a maximum of 12 per cent of current anthropogenic CO_2-C equivalent (CO_2-C_e) emissions (that is, 1.8Pg CO_2-C_e per year of the 15.4Pg CO_2-C_e emitted annually)'. Over the course of a century, they add, the total net offset from biochar would be 130Pg CO_2e.

These are enormous numbers, and not everyone is 100 per cent convinced by biochar's potential. A report by the UK Biochar Research Centre, commissioned by the British environment and energy ministries, endorses biochar's climate mitigation possibilities, but does not find it generally economic for use in Britain until those possibilities can be monetized. It also points out that its carbon neutrality depends on the way the feedstock has been produced, and warns of other questions, such as the import of contaminants onto the land via the feedstock, uncertainties regarding residence times and biological qualities, and other unknowns (Shackley and Sohi, 2010, pp14–15, 29). In November 2010 a report for the Natural Resources Defense Council in the United States commented that: '[F]or a time, biochar seemed to enjoy status as a miracle cure to the global climate challenge. However, there remains a great deal of uncertainty with respect to the environmental and economic performance of different biochar production … [E]stimates of the potential for biochar production and carbon sequestration remain highly uncertain and largely premature at this time' (Brick, 2010, piv).

The uncertainty is increased by the fact that there is not yet much research on the nutrient properties of biochar; and as Chan and Xu (2009, p68) point out, the composition of biochar will depend not only on the feedstock but also the way in which pyrolysis took place. Negative productivity responses to biochar have been reported, apparently as a result of pH increases. Also, the use of solid wastes to produce biochar could introduce toxic materials into the soil (Chan and Xu, pp73, 79–80). It has also been found that, while the use of biochar has the potential to reduce N_2O emissions as well as sequestering carbon, under certain circumstances and using certain feedstock the opposite may apply (Van Zwieten et al, 2009, pp228–229).Woolf et al themselves state that: 'Substantial uncertainties exist … regarding the impact, capacity and sustainability of biochar at the global level.' Glover (2009, p375) warns that, for its potential to be realized, its properties 'need to be verified and taken to market [and] presented to customers as a viable and cost-effective offer … [T]here is potential for over-enthusiastic exploitation that could damage or delay the medium-term commercial prospects for biochar.'

Indeed, there is already some opposition to the use of biochar amongst environmental pressure groups, who see it as a biofuel – which is not its main purpose. Thus Ernsting and Smolker (2009), of the NGO Biofuelwatch, heavily criticize biochar, largely on the grounds that its exact dynamics are as yet uncertain (although no one really denies this). They conclude that: 'When large scale energy crops are required, as would certainly be the case if biochar is adopted as a strategy to reduce atmospheric greenhouse gas levels, emissions from land use change become very concerning. Clearing of forests or grasslands to make way for energy crop monoculture results in large quantities of emissions, reduces future sink capacity and causes further collapse of ecosystems and the biodiversity on which we depend for climate regulation.'

But biochar is not really being advocated as a biofuel. In fact Woolf et al (2010) argue that its greatest climate-change mitigation potential would arise from its being used in soils, *not* for energy generation. In any case, serious supporters of biochar are well aware of the uncertainties and clearly acknowledge them. Many of the potential drawbacks and unknowns, as well as advantages, of biochar have been set out in a comprehensive recent survey co-edited by one of the leading proponents of biochar, Johannes Lehmann of Cornell University (Lehmann and Joseph, 2009), from which much of this section has been taken.

Moreover, biochar has an additional possibility; planting of feedstock for this purpose might – at a stretch – be considered an afforestation and reforestation (A&R) activity under article 3.3 of the Kyoto Protocol, rather than as an agricultural activity (which would come under article 3.4). Agriculture could thus be brought under the Kyoto Protocol by the back door. (The position of agriculture in relation to Kyoto is discussed in Chapter 4.) It is arguable whether this would really be possible under the Protocol; moreover, the use of A&R provisions to plant feedstock might raise some of the same difficulties as plantation for biofuel.

Glover (2009) is surely right to ask that biochar is not oversold, but neither should it be wilfully misunderstood. The use of pyrolysis may be an important tactic in the use of agriculture to mitigate climate change, and should not be dismissed. However, much research remains to be done; moreover, the effects of biochar will always be site-specific.

From Theory to Reality

This chapter asked whether agriculture in developing countries can usefully mitigate climate change. As we have seen, in theory it can. It probably could not alter the carbon flux enough to reduce the atmospheric carbon pool, but it could offset a substantial amount of the increase. Also, because it is found in the areas where the land may turn from sink to source, on-farm carbon in the developing world should be seen as inherently more valuable than that in higher latitudes. If this is the case, perhaps carbon on tropical farms should not be regarded simply as a fungible unit of exchange on the international market, but should be seen as a 'premium' product. In fact, the potential for on-farm carbon sinks to earn money for the poor looks high.

However, this chapter also raised some difficult questions. One was the interaction beween soil carbon sequestration and two other significant greenhouse gases, CH_4 and N_2O. There is also an urgent need to reduce emissions of these from agriculture – to the extent that much talk of mitigating emissions from agriculture has little to do with carbon at all. No carbon sequestration intervention makes sense if its benefits are outweighed by increases in non-CO_2 GHG emissions that the intervention has itself caused. Thus pasture improvement could sequester large amounts of CO_2, but if it does so through increased use of N fertilizer, it will be necessary to ensure that there is really a net gain. This will be especially so if the improved pasture has a higher carrying capacity, causing the farmer to graze more cattle – and increases CH_4 emissions from enteric fermentation. This is an issue that will recur in the later chapters of this book, in which soil carbon sequestration practices are discussed in more detail in the context of a field survey.

Another theme that has run through this chapter has been uncertainty. Estimates of even theoretical soil-sink potential are approximate, and are subject to non-biophysical drivers such as carbon prices and adoption rates. Moreover, some of the physical science is uncertain. Not least of the unknowns is the enormous one of whether soil erosion is a global sink or source. Biochar, too, raises as many questions as it answers. To some extent, these questions are unanswered not because they are difficult, but because no one has hitherto needed to answer them. However, some *are* hard – especially those around soil erosion, control of which also poses more non-biophysical challenges than might be supposed. There will be more on this in later chapters.

The next chapter reviews more basic potential constraints to monetizing CO_2 mitigation in agriculture. The first is, quite simply, whether there is a basis for monetizing carbon. If there is, is it even ethical? And should lower-income countries be involved?

Notes

1 OECD is the Organisation for Economic Co-operation and Development and its membership is often used to define developed countries; by this criterion, there are now 33. EIT denotes Economies in Transition, meaning those that have been moving to a free-market economy since 1991; however, the distinction between this group and the other categories can be blurred.

2 FAO (2002b) gives a breakdown by region which is very roughly comparable to the one quoted here. Niles et al's figures have been used because they cover a more convenient period; also, the FAO figures quote a range, which is very

broad. FAO has factored in reduced tillage regimes but says it has not included reduced emissions from the lower erodibility that would be expected. Niles et al do not appear to either, but the amounts involved could be significant (although this is open to dispute – a subject discussed later in this chapter).

3 It is not always clear whether FAO has included some indirect emissions from agriculture through, for example, CO_2 emissions from fossil-fuel use in manufacture and transport of fertilizers, mechanization in agriculture and pumping of irrigation water. Also, although FAO does discuss worldwide land degradation due to erosion and salinization, it does not appear to include loss of soil carbon due to these processes in its discussion of emissions from agriculture.

4 US-EPA uses a geographical classification called China/CPA (Centrally Planned Asia), which besides China includes Cambodia, Laos, Mongolia, North Korea and Vietnam.

3

Three Questions on Carbon Economics

> *I have never known much good done by those who affected to trade for the public good. It is an affectation, indeed, not very common among merchants, and very few words need to be employed in dissuading them from it.* (Adam Smith, *The Wealth of Nations*, 1776)

So far, this book has looked at, first, the impact climate change might have on agriculture, and second, how agriculture might have the potential to mitigate climate change through accumulation of carbon. These have been treated as scientific questions. However, this book is reviewing not just the theoretical mitigation potential, but whether it can be monetized for the benefit of farmers in the developing world.

This second question can conveniently be broken down into three separate ones:

- First, can the value of carbon be translated into monetary terms? On the face of it, it is an abstract concept that cannot be bought and sold.
- Second, if it is possible, is it ethical – that is, should we be buying and selling a common resource on which we all depend? Or is it like – in effect – privatizing the atmosphere?
- Third, should poorer countries really be paid to clear up the mess that richer ones have created in the atmosphere? Is it to their advantage?

Regarding the first question, a theoretical economic basis must be established for the trading of carbon – indeed, any abstract environmental good – in some form. This is difficult, as most economic paradigms are dominated by the theory that goods are worth what people will pay for them – a paradigm sometimes expressed in the theory of marginal utility; crudely stated, the relationship between how much of a commodity we have or have access to, and how much we would pay to add to that stock.

On this basis, there is no incentive to pay for something that is free. This leaves little space for constructing an artificial value around a hitherto untraded commodity. It will be suggested that the principles of economics as applied to natural resources are subject to change, and that the operation of marginal utility should not be seen as set in stone. After all, it is a product of the later 19th century; in the 18th century, the notion that carbon had fungible value would have been clear enough, had science then demonstrated its significance. But in the 19th century, views about what did or did not create value changed – not necessarily because economics is socially constructed, but because the world it was examining had changed during the process of industrialization. There is nothing to prevent a further evolution that takes us full circle, and the growth of market environmentalism and ecological economics will be presented in that light.

The second question – whether emissions reductions *should* be traded – is sometimes presented as a practical matter – a utilitarian decision as to what will most reduce emissions. However, attitudes to carbon trading are often coloured by ideology; indeed many would argue that a utilitarian approach to emissions is ideological in itself, and that there is an underlying ethical debate. Should one be allowed, in effect, to buy the right to pollute? Such a transaction implies that there can be such a right; perhaps there cannot be. Indeed, is it ethical to value natural resources at all, when such valuation implies that their exhaustion might be an option?

Finally, how do these two questions relate to the third, regarding North–South issues?

With all three questions, it is too easy to become bogged down in endless ideological arguments. But it is equally easy to insist on the utility of policy over its equity, forgetting that a workable policy must have both. This chapter, unlike the rest of this book, reviews broader climate-change questions that do not relate solely to agriculture. However, it is important background for Chapter 4, which describes how agriculture might fit into the growing carbon market in theory – that is to say, which trading instruments might be suitable; and Chapter 5, which looks at how agriculture might fit into the market in practice – including the thorny questions of monitoring and verification.

It should be noted that, for brevity, some important areas are discussed only briefly in this chapter (for example, the question of whether it is possible to substitute human for natural capital; and the so-called 'tragedy of the commons'). These areas have a literature of their own; the purpose of this chapter is to provide an introduction.

The New Physiocrats

Why have we hitherto not priced the environment? Former US Vice-President Al Gore has described how, as a child, he listened as an apparently rational man explained to Gore's father that he could turn lead into gold. Gore suspected his father wanted him to hear one of the last alchemists in history. But, he argues, there is little more logic in the way classical economics prices our use of the environment. 'When the number of good things produced with each unit of labor, raw materials, and capital go up ... then productivity is said to increase. But what if [this] results not only in the production of good things but also in an even larger increase in the number of bad things?' (Gore, 1992, pp187–188). They may even be accounted as a positive rather than a negative input to the economy. Gore uses the example of the *Exxon Valdez* oil spill; the clean-up efforts actually contributed to GNP (ibid.). Others have noted this double anomaly. Pearce et al (1989), in the introduction to their influential *Blueprint for a Green Economy,* note that if pollution damages health, the resulting health-care expenditures are counted in GNP and our standard of living has, supposedly, risen.

Further anomalies are associated with resource use. Costanza et al (1997a, p121) use a hypothetical example of a developing country that obtains 6 per cent of GNP from timber exports; if 2 per cent comes from sustainable yield exploitation and the balance from deforestation, then growth has been vastly overestimated: '[I]n an economy whose conventional GNP was growing at 3 per cent, a 4 per cent reduction is the difference between growth and decline,' they state. Were accounting for GNP to be modified accordingly, many such examples would be uncovered. Blaikie (1985, p18) reported that, in 1976, fertilizer use to offset soil erosion in North America was the equivalent of five extra gallons of fuel per acre. As, under conventional accounting, that extra fertilizer production adds to GNP, it may be that productivity through intensive farming is contributing vastly less to GNP than usually thought, and in some cases might even be a negative.

The valuation of commodities has also long puzzled economists. In fact, as J. K. Galbraith has pointed out (1987, p13), it puzzled Aristotle. 'Well into the 19th century, economic writers would still be struggling with the ... difference between value in use and value in exchange – with the fact that bread and clean drinking-water are useful and relatively cheap while silks and diamonds are much less useful and decidedly more expensive. Surely there is, or was, something ethically perverse here,' says Galbraith. 'It would be considered a major advance in economics when this problem was finally solved.'

By the end of the 19th century, however, most economists seem to have felt that it was solved, through application of the law of marginal utility. The recent focus on public goods, such as the capacity to absorb carbon emissions, suggests that this may now evolve.

The law of marginal utility took shape in line with the Industrial Revolution. One of those who led the way was Adam Smith. In *The Wealth of Nations* (1776), he 'solved' the problem of use vs. exchange values simply by setting the value in use aside, and defining value by exchange value on the basis that something was worth the amount of labour for which it could be exchanged (Galbraith, 1987, pp65–66). Hitherto, one might have quoted a 'just price'; St Thomas Aquinas, for example, thought that 'to sell dearer or to buy cheaper than a thing is worth is in itself unjust and unlawful', but never stated how such a price should be fixed, aside from saying that it should be left to God (op. cit., pp25–26).

By Smith's time, things had moved on. The dominant thesis in some countries was mercantilism – crudely put, that a country should practise protectionism in order to accumulate precious metals. But it had been challenged, at least in France, by the Physiocrats, the leader of whom was the surgeon François Quesnay. The Physiocrats held that only agriculture produced a surplus over the cost of production (Raphael, 1997, p24). 'The soil is the only source of wealth,' wrote Quesnay. 'Those who buy the products of industry pay for the costs, the manual work, and the merchants' gain; but these commodities do not produce any revenue beyond that' (Quesnay, *Œuvres économiques et philosophiques*, quoted in Stark, 1944, p15).

Adam Smith spent most of 1766 in Paris and knew Quesnay well. Smith had a high regard for him, and thought the Physiocrats' views 'generous and liberal' (*The Wealth of Nations,* quoted in Raphael, 1997, p75). But he did not share them. For Smith, the sources of wealth included labour and the skill with which it was deployed.

This advance in thinking must have reflected at least in part the differences between the worlds in which the two men lived, friends though they were. Quesnay was physician to Louis XV and Madame de Pompadour, and had apartments in the Palace of Versailles – indeed Smith sometimes met him there (Raphael, 1997, p24). The Physiocrats' perspective on industry was understandably limited; for Quesnay's colleague Cantillon, for example, the prototype of an industrialist was a hatter, and like hatters, all tradesmen were ultimately supported by landowners (Stark, 1944, p18). Smith, by contrast, was from the cradle of the Industrial Revolution and his lectures show that he had either visited modern factories, or knew people who had done so (Raphael, 1985, p46).

Neither Smith nor the Physiocrats need have been right or wrong; rather, they were analysing different worlds. This does not mean that economics is not an objective discipline. But as the world changes, economists' perceptions of it must also do so. Economics will reach different conclusions at different stages of economic development, and this in no way invalidates it as an objective discipline. This view is not new; indeed Galbraith (1987, pp93–94) ascribes its beginnings to the German-born economist Georg Friedrich List, who was born, just, in Smith's lifetime. 'Do economists … seek and find eternal verities as, say, chemists and physicists do? Or are the institutions with which economics deals in a constant process of transformation to which the subject, and more particularly the policies it urges, must be in a similarly constant process of accommodation?' asks Galbraith (ibid.). It seems he takes the second view. Werner Stark (1944, pp74–75), in his short and elegant *The History of Economics in its Relation to Social Development*, takes a related line, seeing economics as allying itself with different disciplines according to the era. Thus in the time of mercantilism it was related to jurisprudence; in that of the Physiocrats, with philosophy and theology; and in the 19th century with history. Stark would presumably have seen no reason why economics should not reflect our own era by binding itself to environmentalism, and this is coming to pass.

We should therefore not hesitate to challenge assumptions arising from the work of Smith and his successors. Indeed, the new environmentally driven economics rejects two implications of Smith's work. One is the assumption that human-made capital – the products of labour – can substitute for natural capital; this would have been widely accepted from Smith's day nearly to our own, but the Physiocrats would not have accepted it, and modern market environmentalism does not entirely do so. The sub-discipline of ecological economics, discussed later in this chapter, does not accept it at all.

The second assumption that dominated the 19th and 20th centuries, and which also must now be laid aside, is that the value of a commodity is its exchange value – a view that was later refined into the theory of marginal utility. This has its roots in the later 19th-century work of Alfred Marshall, who saw an interplay between supply and demand, and William Stanley Jevons. In essence, if a commodity costs 65p and we get £1-worth of pleasure or utility out of it, we will buy it; but that pleasure or utility may decline the more times we buy it, and at some point the margin between the satisfaction obtained and the price will disappear.

In theory, there is no reason at all why we should not apply this principle to carbon emissions. If stabilizing the atmospheric carbon pool at 450ppm costs more than the benefit we expect from it, we will not do it, but stabilizing emissions at 750ppm might be worthwhile. This principle was in fact applied to natural resources by Hotelling (1931), in what is sometimes seen as the seminal article on this subject.

In its classical form, however, the theory does not work for natural resources, for two reasons. The first is simply that the marginal utility of emissions regulation will differ between those who are more or less vulnerable to climate change. This is well expressed by Azar and Schneider (2003). What, they ask, if emissions primarily effect poor countries such as Bangladesh and Mali? '[I]n this scenario … rich countries emit, poor countries get hurt and economics suggest that this is optimal since the rich have the potential to compensate the poor. It would be unfortunate if economics would be the tool that some use to justify, without closer scrutiny, such an outcome' (op. cit., p330). Moreover the rich will not necessarily compensate the poor just because they can do so.

The second reason why marginal utility should not be applied to carbon emissions in so simple a sense is that, like many natural resources, the disposal facility of the atmosphere is an open-access resource, and does not charge the full cost to the emitter (or, to put it the way it is often expressed, that cost becomes an *externality*; that is, instead of being internalized in the transaction, the cost is met elsewhere or by others). The problem with such open-access resources was laid down by Garrett Hardin in a very well-known paper, 'The tragedy of the commons', in 1968:

> *Picture a pasture open to all. It is to be expected that each herdsman will try to keep as many cattle as possible on the commons. … As a rational being, each herdsman … asks, "What is the utility to me of adding one more animal to my herd?" This utility has one negative and one positive component. … Since the herdsman receives all the proceeds from the sale of the additional animal, the positive utility is*

> *nearly +1. ... Since, however, the effects of overgrazing are shared by all the herdsmen, the negative utility for any particular decision-making herdsman is only a fraction of –1.* (Hardin, 1968, p1248)

In other words, the herder who puts an additional animal on common land will reap all the benefits of doing so, while the costs of putting it there will be shared across all the land users. The rational herdsman, argues Hardin, will therefore add another animal, and another – and so will everyone else. 'Therein,' he states, 'is the tragedy.'

The same principle can be applied to the disposal function of the atmosphere. As Costanza et al (1997a, pp167–168) state, a nation that internalizes environmental and social costs will be at a disadvantage when trading with one that does not, and its firms will be driven out of business. Their response is to see a conflict between responsible environmental economics and free trade. External costs, they argue, are now so important that they should take precedence over the latter.

Hardin's argument has a flaw, however; it stresses the utility of selfish resource use but it ignores the marginal utility of *cooperation* in resource use – which, if it results in the preservation of the resource, will ultimately be greater. As Sandford (1983) argues in his study of range management, there is an alternative form of rational self-interest, and that is for individual groups to come together in agreements that will preserve the resource for all. Historically, this is in fact what pastoralists have done, although this has not always been obvious to outsiders. This does not negate the law of marginal utility in resource use, but does demonstrate that it may be applied in different ways. As Stark wrote over 60 years ago, some considered 'the law of marginal utility ... a piece of eternal truth ... [but] any orthodoxy must one day come into conflict with facts and succumb' (Stark, 1944, p57, fn). It has not exactly succumbed, but must be applied in different ways when the available facts change.

The first question for this chapter was whether we can assign carbon a monetary value. In conceptual terms, at least, we can. The theory of marginal utility can accommodate the pricing of a common resource if we update it by adapting the emphasis of economics for our times, just as Smith, Marshall and Jevons did for theirs. We can do this, in effect, by recognizing the marginal utility of *cooperation* in resource use when we assign value to carbon. What must be reviewed next in this discussion is whether such values can be assigned to carbon in practice.

Costing Carbon: Shadow Pricing?

There are two ways in which a price can be fixed for carbon. The first is to attempt to assign a shadow price to a tonne of carbon emissions, and price permission to emit accordingly. The second is to let the market do it, with countries prepared to pay whatever it is worth to them.

In either case, it will have to be decided how much can be emitted, and how the initial permits should be distributed; that is discussed later on. However, the question of whether a shadow price can really be calculated is of great importance, because if it cannot, then certain types of instrument for emissions limitation – in particular, a carbon tax – are not practical. How can we know what the tax rate should be? Markandya (1991, p55) points out that too low a tax could result in too low a level of emissions reductions, with serious environmental consequences. Given countries' reluctance to make commitments at all, this seems all too likely. Without a meaningful shadow price, negotiating international carbon-tax differentials would be virtually impossible.

Can a shadow price be calculated for a tonne of carbon? Doing so requires estimation of externalities, including damage from increased weather events, loss of food production, health impacts such as malaria and parasitic diseases and much more besides. In theory, it is possible. Costanza et al (1997a, p203) lay out the principles for fixing optimal charges for emission of a pollutant, which CO_2 could be seen as in this context. First, one must estimate the relation between emission levels and ambient concentrations; then estimate the biological damage associated with those concentrations; and finally assign economic values to given levels of biological damage. However, they comment that early assessments of how this might be done for individual pollutants 'appear to have been overly optimistic, now that ecological economists are learning actually to develop transdisciplinary working relationships with physical scientists' (ibid.). Multiple emissions sites, species and interactions between different pollutants will, they say, complicate things further.

For CO_2 these problems are about as complex as could be. To take just one example, calculating loss of rangeland productivity would require modelling future grazing patterns and their effect on the balance of plants with different photosynthetic pathways, as this will affect response to elevated carbon dioxide and thus the temperature at which biomass starts to decrease. A similar range of challenges must be met for every ecosystem if any shadow price is to be accurate. So it is not surprising that estimates so far for the externalities of a tonne of CO_2 equivalent (CO_2e) cover a very wide range. Pretty and Ball (2001, p20) have suggested $20 to $95 per tonne, but it can be even wider. Dessler and Parson (2006, p121) report that the technique has so far been to divide economic sectors, make estimates of their climate sensitivity and the value of probable impacts, and then add them up. This approach – which, they say, has often been 'highly simplified and judgmental, even arbitrary' – has yielded estimates of 1 per cent to 2.5 per cent loss of GDP for a doubling of CO_2 concentrations to 550ppm. 'Attempts to assess comprehensive damages ... in ... marginal costs per tonne of emissions have spanned a wide range, from $1–2 per tonne carbon through more than $300,' they report. 'Most estimates lie between $10 and $100, and depend strongly on the discount rate used: studies using discount rates around 5 per cent give estimates that fall near the low end of this range, while those using rates around 1 to 2 per cent fall near the high end.' The reference to discounting in itself raises difficult questions of inter-generational equity, making the time horizons used for these calculations almost impossible to fix.

Moreover, in the case of soil carbon in agriculture, its loss has externalities unrelated to climate change (for example, loss of SOM can increase incidence of flooding); this further complicates accounting. Izac (1997, p273) sets out a formula for calculating the shadow price of soil carbon – but from the farmer's point of view, not everyone else's. She states: '[T]he various ecological, environmental and agricultural functions of soil carbon [must] first be identified. The shadow price, or economic value to society of soil carbon in tropical countries, can then be directly related to the physical, chemical and biological functions of soil carbon in tropical ecosystems.' But should these functions be included in shadow pricing intended purely for emissions limitation? That would, in effect, be hidden taxation. Besides, her calculations are based in part on the cost of substituting inputs, including inorganic fertilizer, for soil carbon as the latter declines (ibid.). This implies the substitution of manufactured for natural capital, and, as stated earlier, many argue that this cannot be done. Moreover, as was mentioned earlier, the cost of repairing environmental damage – e.g. manufacture of the extra fertilizer – would be regarded as contributing to GDP, an example of the perverse effect of a pollutant on the economy as we have hitherto viewed it.

In fact, while it may be possible to calculate shadow prices for some pollutants, in the case of GHGs the variables (and the shared assumptions required) may just be too numerous. Without such a shadow price, it is hard to see how any basis can be agreed for taxing carbon emissions, either within or between nations. But then, what is calculable in theory is often not in practice. The 19th-century economist Alfred Marshall so loved mathematics that he read mathematical treatises in front of the fireplace, but feared economists would 'calculate themselves into irrelevancy' (Buchholz, 1999, pp156–157). There is a risk that attempting to calculate shadow prices for CO_2e would have that effect.

Costing Carbon: Using the Market

The alternative to establishing a shadow price is to decide how much we can afford to emit, and then allocate permits up to this level. These permits can be bought and sold. The price per tonne of CO_2 will then be set by the market, but a 'wrong' price will not adversely affect emissions, because the amount that we emit has been capped.

The use of the market has the enormous advantage that it reduces the cost of emissions reductions to the polluter. An example of this might be a UK power generator, running a modern plant with relatively low emissions, but still forced to offset some of them, perhaps through rising demand. To further improve the already efficient power station might cost (say) £25 per tonne/day of emissions reductions achieved. However, the operator can instead invest the same money in an ageing Chinese or Indian coal-burning plant and produce several tonnes' saving for the same price. So emissions reductions will cost less, because the polluter will be able to find the cheapest place to bring them about. They might also take place in countries that would not otherwise make improvements, either because they are already emitting below their quotas, or because they are developing countries and do not have them. Tietenberg (1990, quoted in Markandya, 1991, pp60–61) made an early estimate of the reduction in per-tonne mitigation costs in comparison with direct regulation. If emissions were reduced by 20 per cent over then current levels – that is, about a billion tonnes of carbon – the cost would have been (at about £20 to £30 a tonne) £20 to £30 billion using a trading system, but £100 to £150 billion using relatively inefficient regulation.

In the long term, the savings might be greater, as a tradeable permit system could encourage technical progress in emissions reduction; Tietenberg argued that this had already proved to be the case with certain pollutants.

Of course, this would apply in theory to direct regulation as well, but as the permitted emissions levels would probably not be the same in every country, exporters might just shift markets instead. Something of this sort could be seen when, in the wake of Ralph Nader's campaign on vehicle safety, the United States introduced difficult crash safety regulations in the 1970s; some non-US manufacturers simply withdrew, especially from California, where stringent emissions restrictions had also been introduced.

The advantage of tradeable permits is that although the initial permit allocation might still vary from one country to another, emissions permits would move around, creating a global emissions system. As Costanza et al (1997a, p41) point out, a systemic approach is better. They use the example of bioreserves, which may simply move the reduction in genetic diversity elsewhere. Soil carbon is comparable; under a regulatory system, if intensive cereals farming runs emissions levels too near the sectoral quota for agriculture in Kansas, the business may simply plough up steppeland in the developing world instead.

At the beginning of this chapter there were three questions. The first was whether carbon could be priced in monetary terms. Fixing shadow prices is only possible in a very theoretical sense, but what we can do is decide how much we can safely emit, allocate the relevant permits and then allow them to be traded. In this way, the market will sort out the price but will do it with reference to carbon's 'real' value, because the number of permits issued has been tied to that. So the answer to the first question – whether we can price carbon – is yes. But it may need to be done by the market, not by regulation or taxation.

A Price for Everything?

The second question, however, is whether such monetary values *should* be attached to carbon. Can we pay for the right to destroy nature? There is a debate, often with mystic and quasi-religious overtones, between those who say that nature is sacrosanct, and those who think it is not, and is subject to exploitation for the benefit of human beings, provided that this is done sustainably.

This is not a new discussion and is sometimes traced to the differing views of the 19th-century nature enthusiast John Muir, founder of the Sierra Club, and Gifford Pinchot, the founder of the United States Forest

Service (USFS). Thus Elliott (1996, pp5–8) sees Muir as the father of preservationism, the chief concern of which was and remains the protection of forests, whereas Pinchot was a proponent of sustainable management: 'The central thing for which conservation stands is to make this country the best possible place to live in, both for us and our descendants. It stands against the waste of natural resources ... and most of all it stands for an equal opportunity of every American citizen to get his fair share of benefit from these resources, both now and hereafter ...' (Pinchot, 1901, quoted Elliott, 1996, p6). The Pinchot view prevailed, perhaps because it was supported by the then president, Theodore Roosevelt, an avid big-game hunter (Adams, 2001, p27). Elliott (1996, pp7–8) sees the descendants of Muir as having split into preservationists and ecocentrists, with the latter having completely rejected an anthropocentric approach to nature: 'The concern,' he explains, 'is to protect species, populations, habitats and ecosystems wherever they are situated and irrespective of their value to humans' (Eliott, 1996, p8).

Just how far apart the two sides have now grown is made clear by M. Lewis (1992), who sub-classifies radical environmentalism into antihumanist anarchism, primitivism, humanistic eco-anarchism, eco-Marxism and eco-feminism. He describes these in some detail, but the labels probably convey enough for this discussion. What is important is their distance, as seen by Lewis, from any type of mainstream opinion (his book is titled *Green Delusions*). For example, he claims that many adherents of radical eco-feminism are 'actively reviving the goddess-centred cults that they believed once allowed humans to live in harmony with nature. Several leaders have proclaimed themselves witches and formed covens devoted to the ancient faith of Wicca' (op. cit., p34). Social progress, argues Lewis, demands broad inclusion, whereas 'radicalism excludes all persons judged sinful – or, in the current jargon, politically incorrect' (op. cit., p249). Lewis has an agenda too. *Green Delusions* is a polemic. But it is true that worshipping Wicca is unlikely to get others out of their sport utility vehicles by any obvious means.

However, not all opponents of market environmentalism are so easily dismissed. Adams (2001, p105) argues that it assumes the substitutability of human-made capital for natural capital, and indeed Izac (1997) made exactly that assumption in her paper on the value of soil capital, discussed above; but that assumption is not universally shared, any more than it was in the days of the Physiocrats. Moreover market environmentalism does not confront the threat that economic growth will cause ecological systems, and eventually the economy, to collapse altogether. Adams admits that calls for zero growth can be naïve and easy to caricature, but warns that some are not so easily ignored – in particular the 'impossibility theorem', which simply points out that a high-consumption Western-style economy is just not possible for everyone on the planet for long (Adams, 2001, pp106–107).

However, acceptance of physical limits to growth should not automatically preclude valuation of the environment. Rather, it may simply mean a different attitude to substitutability, and a stronger appreciation of the limitations of the biosphere. This can be seen in the field of ecological economics, which is something distinct from 'ordinary' environmental economics, seeking as it does to be a holistic study of the ecology and the economy (op. cit., p123).

In ecological economics, certainly, it is recognized that there are limits to growth. Costanza et al (1997a, pp57–61) discuss the ideas of Nicholas Georgescu-Roegen. Put simply, he pointed out that the available energy in any given system could only decline – the entropy law, or second law of dynamics. This states that some energy sources are more concentrated than others, but that this concentration is ultimately dispersed. An example might be a tankful of petrol. It will disperse anyway through evaporation but will do so more quickly if used in an engine. Something similar happens when an egg is boiled and then put in a pan of cold water to cool; the egg cools and the water warms up. The amount of energy (heat) there is the same (apart of course from losses to the atmosphere, the metal of the pan, etc.), but it has been evened out over the system. So in time, everything within a closed system will be the same temperature, and the energy unleashed by its dispersal will no longer be available. Earth, it is argued, is a closed system, sooner or later the energy concentrated in certain points or substances is dispersed within the system, and can no longer be used through its sudden release.

True, the earth is *not* a completely closed system, as critics of Georgescu-Roegen have pointed out; new energy accumulates from the sun as it encourages the growth of plant material, and this will – if left alone – become fossil fuel. But although we can mine fossil fuels, we can't use tomorrow's sunshine; in theory at least, we must slow our resource use to replacement rate. So he is not wrong in principle.

The utility of entropy in valuing the environment is limited. As Costanza et al (1997a, p57) state, it doesn't tell us how quickly we need to make that transition to replacement rate. 'In this sense,' they continue, 'we simply need to look at resource constraints as well as the ability of the global system to absorb carbon dioxide and other greenhouse gases. ... The entropy law, however, does provide a strong bass beat to the sirens being sounded by scientists studying climate change, biodiversity loss, and soil degradation.'

It also provides an interesting alternative model to mainstream environmental economics, accepting limits to growth, rejecting substitutability, and yet still permitting the valuation of environmental resources, including atmospheric absorptive capacity. Indeed Costanza et al (1997b), in a controversial paper in *Nature*, went so far as to value – in the words of the title – 'the world's ecosystem services and natural capital'. They came up with an annual value of between $16 trillion and $54 trillion, with an average of $33 trillion, being 1.8 times global GNP (op. cit., p259). Some biomes, they admitted, were not counted (including cropland – a big omission), and there were substantial uncertainties that were clearly acknowledged in the paper, but they regard their estimates as a useful starting point.

The paper drew mixed reactions. Turner et al (1998, p62) commented that the paper had 'engaged environmental scientists and policy makers, but the global, biome scale economic value calculations risk ridicule from both scientists and economists'. They went on to raise a number of problems with the $33 trillion figure. This is not surprising; the difficulty of calculating a shadow price for an ecosystem service was demonstrated regarding carbon earlier in this chapter. Some responses to Costanza et al's paper took a sharper tone. Norgaard and Bode (1998, p37), while acknowledging that the exercise was thought-provoking, wanted to know: 'Will ecological economists bring us the value of God next? And will this be the end of history for economic valuation? Or, now that we know the exchange value of the earth, we wondered with whom we might exchange it and what we might be able to do with the money, *sans* earth.' Rees (1998, p52) puts it a little differently: '[A]t the present critical stage of world development, we must regard many of nature's services as we would an expensive yacht. If we have to ask the price, we probably can't afford it.'

Costanza et al, however, regarded the exercise as both ethical and necessary. They acknowledged the argument that ecosystems should be protected for purely moral reasons and could not be explicitly valued. But they added that there were equally compelling arguments to the contrary,

'... for example the moral argument that no one should go hungry. ... So, although ecosystem valuation is certainly difficult and fraught with uncertainties, one choice we do not have is whether or not to do it' (Costanza et al, 1997b, p255).

What they do *not* say is that such services should not only be valued, but traded. Ecological economists might not wish to take the commoditization of carbon quite that far. Yet they accept that ecosystem services are frequently undervalued because they are outside the market. So the logic of performing such a valuation implies that such services *could* somehow be incorporated into the world trading system.

Indeed, market environmentalists might argue that true morality lies in doing what must be done to control atmospheric carbon levels – including selling emissions permits. This point was well made by David Pearce, who, as an adviser to Britain's Department of the Environment, co-authored the report that became *Blueprint for a Green Economy* in 1989. In an interview not long before his death in 2005, he was asked if he could justify putting a price on the environment. He replied that he lived 'in the real world of real policy, I look at the forces that destroy nature, and I try to use those same forces to conserve nature'. As an obituarist in Britain's *Independent* (17 September 2005) put it: 'To many people, the idea of pricing the environment seemed immoral. To Pearce, letting it be used for free was worse.'

Is There a Market?

The first question at the start of this chapter was whether there was any theoretical basis for buying and selling carbon, and the answer was yes. The second was whether it *should* be bought and sold, and the answer has also been yes. The third question was whether the developed countries that have mainly caused climate change should be paying other countries for mitigation – indeed, whether mitigation should really be expected from developing countries at all. To answer this, it is necessary to briefly review the debate and negotiations over carbon trading.

The central problem in setting up some sort of carbon market is in allocating the 'permits to pollute'; until this is done, no one can buy or sell any. The problem this presents epitomizes the central problem of all environmental policymaking, summed up by Adger et al (2002, p1097):

> *[H]owever efficient or effective a decision, things tend to unravel if it seriously compromises equity or legitimacy. In a similar vein, the supposed equity or legitimacy of a decision does not seem to justify its lack of efficiency and effectiveness. Therefore, we need to pay simultaneous attention to the four criteria that challenge the problem-solving capabilities of most decisionmakers concerned with environmental governance and with sustainability.*

In short, any environmental policy must be effective; but it must also be just, or it won't be implemented, and therefore won't be effective anyway. The arguments about who should commit themselves to what, and when, in a global climate regime have effectively been about two of these criteria, equity and legitimacy (sought by the developing world) vs. the other two, efficiency and effectiveness.

Negotiations for such a system have taken place under the United Nations Framework Convention on Climate Change (UNFCCC), a product of the 1992 'Earth Summit' in Rio de Janeiro and one of only two agreed there that is legally binding (the other is the Convention on Biological Diversity, or CBD). Brown (1997, p385) defines the UNFCCC's main provisions as follows:

- It acknowledged climate change as a serious problem.
- It agreed that action could not wait until resolution of scientific uncertainties.
- Developed countries should take the lead in reducing emissions.
- Compensation must be paid to developing countries for additional costs of implementing the Convention.
- Developed countries aimed to return to 1990 emission levels by 2000.
- A reporting process was to be established so that it was clear who was emitting what.

The actual implementation of the treaty depends on the Kyoto Protocol of December 1997, at which the parties agreed (among other things) who should do what in terms of cutting emissions during what is known as the 'first commitment period' (to 2012; it is not yet clear what happens then, as a future climate change regime is still being negotiated). Key to subsequent events is the fact that Annex 1 countries (broadly, the developed countries including the so-called transition economies of the old Soviet bloc) agreed to make cuts; the developing countries (everyone else, including rising industrial powers such as China, Brazil and India) did not. It is largely this that prevented the United States from ratifying the Protocol; as a result, it could not enter into force until most other key players had ratified, and this was not until 2005.

However, the developing countries – including the big players such as India and China – have not seen why they should commit themselves when the historic responsibility for elevated atmospheric CO_2 lies elsewhere. The perception has been that the developed countries, having caused the environmental crisis as part of their own industrialization, now want the rest of the world to forgo its own development in order to deal with it. This perception first emerged at the Stockholm conference on environment and development in 1972. It did not express itself as an outright rejection of environmental reform. But as a liberal Brazilian diplomat, João Augusto de Araujo Castro, wrote in the same year, developing countries 'could only share a common responsibility for the preservation of the environment if it was accompanied and paralleled by a corresponding common responsibility for development' (quoted in Araujo Castro, 1998, p39).

The developing countries have a strong case. Baumert and Pershing (2004, p12) explain that cumulative data is now available back to 1850; and that industrialized countries account for a substantially higher percentage of emissions if this is taken into account, than they do on current figures alone. Thus the European Union as a whole was responsible for 16 per cent of current GHG emissions, but 27 per cent on a cumulative basis; while the United Kingdom, the world's first industrial nation, had a cumulative figure nearly three times its current emissions. But China and India's cumulative share was only half their current percentage.

Neither is historic responsibility the only argument. In what has proved to be a seminal paper, Agarwal and Narain (1991, p5) argue that it is quite wrong to equate a tonne of CO_2 from a luxury car in the North with the methane emissions from water buffalo and rice fields that feed the poor. 'Do these people not have a right to live? No effort has been made … to separate the "survival emissions" of the poor, from the "luxury emissions" of the rich,' they say.

Agarwal and Narain's 1991 paper was in fact a response to the World Resources Institute (WRI) annual *World Resources* report for 1990–1991. The editorial board that had overseen the report had been chaired by Indian scientist M. S. Swaminathan, well known for his part in the technology transfer behind the Green Revolution in the 1960s. The cause of Agarwal and Narain's anger was WRI's contention that climate change (WRI used the then-current phrase, global warming) was a responsibility of all emitters, not just the industrialized nations:

> *Three of the six countries that are the largest contributors to the atmosphere's warming potential – the United States, the USSR, Brazil, China, India and Japan – have heavily industrialized economies; three do not. ... Such widespread responsibility for significant greenhouse gas emissions means that any effective agreement to stabilise or reduce these emissions will have to be equally widely based.* (WRI, 1990, p15)

The report went on to warn that if China and India alone were to raise their per capita emissions to the current world average, worldwide emissions would increase by 28 per cent – and if they raised them to the per capita level of a developed country (they used France as an example), they would increase by 68 per cent (op. cit., p17).

Agarwal and Narain's response to this was sharply negative. 'By shifting the onus onto the developing world, it whitewashes the role and responsibility of the West,' they argued, adding that 'Third World environmentalists must not get taken for a ride by this highly partisan "one worldism"' (Agarwal and Narain, 1991, p3).

This statement was not fair. The WRI report had made it very clear that in historical terms, the developed world was to blame, printing cumulative emissions figures for 1950–1987; in these, China ranked 4th, India 9th and Brazil 15th (WRI, 1990, p14). They also showed per capita emissions; true, Brazil is in seventh place because of deforestation (the extent of which Agarwal and Narain dispute) – but India and China do not even figure in the top 50. The WRI report states that 'solutions must … address the need for equity and for industrial development' (WRI, 1990, p3) and adds: 'In the longer term, considerable attention should be devoted to the components of a "global bargain" that will allow an equitable sharing of responsibilities for addressing the problem' (WRI, 1990, p30). In any case, India's own emissions are no longer associated solely with water buffalo, if they ever were; it too has a stake in the consumer economy.

However, while Agarwal and Narain's attack on the WRI was shrill, it also contained interesting perspectives on the ways in which global responsibility for emissions should be divided. Simply put, they state that each country has not just a percentage share of emissions, but also a share – divisible per capita – of the earth's absorptive capacity. Those countries that have not yet emitted beyond their share of that capacity are, they argue, without responsibilities. 'In fact,' they say, 'India can double its total carbon dioxide emissions without threatening the world's climate' (Agarwal and Narain, 1991, p10).

In one key sense, this argument does not work, because every tonne emitted goes into a common atmosphere; as they themselves say (op. cit., p19), free exploitation of a common-property resource 'inevitably leads to ... the well-known "tragedy of the commons"'. The fact that a given tonne emitted came from a country that had not exceeded its quota makes absolutely no difference to its radiative forcing capacity. Neither does whether it came from the back end of a Buick or a buffalo. Such arguments are doubly futile if you are more vulnerable to climate change than everyone else, and, as was stated in Chapter 1, climate change is likely to hit the poor harder than anyone else.

Agarwal and Narain are, however, on much stronger ground when they suggest that a good way of addressing the inequities is through tradeable emissions:

> *Expecting everyone to adhere to a standard pollution limit does not provide any incentive to low-level polluters to keep their pollution levels low. In other words, what the world needs is a system which encourages a country like India to keep its emissions as low as possible and pushes a country like [the] USA to reduce its emissions fast.* (Agarwal and Narain, 1991, pp19–20)

The advantages of a market system have already been discussed (see above). However, there are two interesting points arising from this quote. First, the authors are stressing the incentive the system gives to low emitters to *stay* low, and this has positive implications for the long-term preservation of carbon sinks, in forests as well as agriculture. This is very relevant, given the problems presented by permanence in sinks; this is discussed in the next chapter.

The second point is one that the authors do not mention: a system of tradeable credits may provide some basis for agreement between developed and developing countries – including the latter's acceptance of emissions quotas. And if emissions levels are to be effectively stabilized, then some developing countries at least must reduce their emissions. China, especially, is apparently now the largest national source of global emissions. It is hard to see how an effective climate regime can be built without those countries accepting caps of some kind. Corfee-Morlot and Höhne (2003, p281) argue that 450ppm (the level generally seen as required to limit temperature increases to 2°C) will not be attainable in the long-term without some developing, as well as developed, countries dropping below their baselines by 2020. Indeed, Berk and den Elzen (2001, p473) argue: '[I]f the group of countries adopting quantified commitments after the first commitment period [of Kyoto, ending in 2012] is limited to middle-income developing countries, and this would set a precedent for future extensions of the group of participating countries, stabilization levels of 550 ppmv or lower may be out of reach.' It is pointless arguing about historic responsibility while the water laps through your living room.

However, the ideological gap between developed and developing countries may eventually be closed, perhaps by tradeable emissions quotas but also by other trends. Developing countries do have other incentives to reduce their emissions. Chandler et al (2002) looked at evidence of this in six major developing countries (Brazil, China, India, Mexico, South Africa and Turkey; the extent to which all of these can be classified as developing is debatable). They found that the six between them had carbon emissions roughly 18 per cent lower than they would have, had they not taken measures to increase efficiency over the previous 30 years. The difference amounted to about 300 million tonnes; to put that in perspective, the reductions in *developed* countries by 2010 required by the Kyoto Protocol amount to 392 million tonnes.

These efforts in developing countries have, say Chandler et al, been driven 'not by climate concerns but by imperatives for development, poverty reduction, local environmental protection, and energy security' (Chandler et al, 2002, p52). This process may be even faster than they think. For example they point to Brazil's measures to reduce dependency on imported energy, which have included the substitution of ethanol for petrol, but that programme had faltered badly by the time their paper was published, due to rising ethanol prices, driving many such cars off the road. However, it has since been boosted by the introduction of TotalFlex cars that run on any combination of ethanol and the local petrol (which does already contain some ethanol).

Moreover there is striking evidence in Baumert and Pershing's (2004) figures for carbon intensity (a measure of GHG emissions per $ of GDP).

Between 1990 and 2000, a few countries (including Brazil) saw increases in intensity. But five countries simultaneously experienced intensity declines and GDP increases exceeding 20 per cent. The most striking case was China, where intensity fell 47 per cent while GDP grew 162 per cent. 'It remains to be seen whether these trends are anomalous one-time shifts reflecting particular circumstances ... or whether they suggest the potential for a longer-term decoupling of economic and emissions growth' (Baumert and Pershing, 2004, p8).

If the latter is the case, the major developing countries may be happier with caps in the future. Even now, they may be more open to caps than is realized, provided they are fair. Jung et al (2005) interviewed a range of stakeholders – policymakers, business people, academics and NGO representatives – in a number of Asian countries, researching Asian perspectives on climate change. They found, to their own surprise, that about 70 per cent agreed in principle that developing countries would need to accept emissions commitments, although this was hedged around with references to historic responsibility and the need to differentiate between countries (Jung et al, 2005, pp73, 75).

What Now for the Climate Regime?

So some agreement should be possible, but how do we bring it about? Corfee-Morlot and Höhne (2003, p288) suggest three options for bringing emissions down to the intermediate target they think appropriate for 2020 (which is between 23 per cent and 50 per cent above 1990 levels). The options are as follows:

- Developing countries do not assume any emissions targets; rather, Annex 1 countries tighten theirs.
- Some developing countries accept targets but others do not. The criteria would be based on GDP per capita; say $12,000 in 2010, so that Argentina, the Gulf States, South Korea and Singapore reduce emissions at the same rate as the Annex 1 countries.
- All countries have targets, with per-capita emissions targets. These need not be the same initially, but would converge by 2050 at a rate designed to meet the global emissions targets deemed acceptable. This approach is known as contraction and convergence (C&C).

The first option was never likely to work for reasons already discussed; in the wake of the 2008 recession and with Japan's rejection, in December 2010, of a future Kyoto Protocol, it looks less likely than ever. The second looks better, but would require complicated negotiations and would exclude countries that are low per-capita emitters but high emitters in absolute terms. China is the obvious example. So Annex 1 countries would still have to shoulder a higher percentage of the burden than they are likely to accept.

The third option is both more workable and just. As Berk and den Elzen (2001, p476) point out, GHG emissions have been presented as a pollution problem, but C&C redefines it as a resource-sharing problem, which is what it is. 'Instead of focusing on emissions reductions, this approach considers the atmosphere to be a global common to which each human being, in principle, is equally entitled' (ibid.). This is essentially what Agarwal and Narain (1991, p10) argued, albeit in different form. The beauty of this is that no one could reasonably disagree with it. What country would argue that their citizens had more right than others to the air that they breathe?

Najam et al (2003, p227) list some other advantages: it is a transparent and predictable regime; everyone will know what is expected of them; and every country would have emissions obligations, thus answering the objection of the US and others to the Kyoto regime.

True, Dessler and Parson argue, it could be held that the poorest countries need not take on any commitments at all:

> *Eighty per cent of global carbon emissions are the doing of only 20 countries, which are either rich, highly populated, or both. At the other end of the spectrum, the 48 countries categorized as 'least developed countries' by the UN amounted to a mere 0.5 per cent of global carbon emissions in 1999. Slowing global warming does not require these countries to take on emission commitments, even over the coming decades*. (Dessler and Parson, 2006, p146)

But perhaps they should anyway, because – as Berk and den Elzen point out (2001, p478) – the least-developed countries (LDCs) might end up with surplus emissions allowances. They see this as a chance for them to develop. However, it is better seen as an opportunity for the LDCs to sell emissions allowances that they do not for the moment need. In any case, it is almost certain that the LDCs would not be the only developing countries to end up with 'headroom' under C&C.

Baumert and Pershing (2004, p10) quote per capita emissions for the top 25 emitters in 2000. At 6.8 tonnes CO_2e per capita, Australia, not the US, tops the list; however, the latter is not far behind at 6.6 tonnes. The average for developed countries is 3.9 tonnes (although the EU is collectively well below that). For developing countries it is 0.9 tonnes, and the average for the world is 1.5 tonnes. Were the latter figure to be taken as the median for C&C purposes, Turkey, Mexico, China, Brazil, Indonesia and India would all be left with varying amounts of 'headroom' (in the case of India, 0.5 tonnes per capita). Adding the LDCs to the average would increase it. In fact India could be left in the happy position of being able to sell emissions allowances to fund its industrial development. Moreover, it would become advantageous for a country to augment its allowances by *creating* carbon through the development of sinks. Agriculture would be ideal for this, giving the LDCs in particular an extra chance to participate in the market.

The most compelling argument in favour of contraction and convergence, however, is equity. C&C acknowledges that everyone has a right to an equal share of the atmosphere. In the end, no regime will work if it does not do this. But if it does, all else will follow.

Before anything like this can happen, however, developing countries must accept more responsibility for mitigation. In December 2007 this question threatened to undermine the United Nations Climate Change Conference in Bali – known officially as the 13th Conference of the Parties, or COP13 (although other meetings also take place on such occasions). That COP13 was not derailed by this question was due in part to the last-minute agreement of clause 1bii in the Bali Action Plan (or Roadmap, as it is widely known), which sets out the agenda for negotiations on a post-2012 climate regime. The clause calls for the negotiations to consider: 'nationally appropriate mitigation actions by developing country Parties in the context of sustainable development, supported and enabled by technology, financing and capacity building, in a measurable, reportable and verifiable manner' (UNFCCC, 2007, clause 1bii). Paragraph 1 of the Roadmap includes other important provisions, for example on adaptation and on avoided deforestation, and these have a bearing on clause 1bii; but it was the clause itself that enabled the Roadmap to be agreed.

The key to its importance is not the fact that developing countries have agreed to mitigate their emissions; under the UNFCCC, everyone has always agreed on that in principle. The key is the demand for actions that are 'measurable, reportable and verifiable' (MRV).

Much drama attended the agreement of clause 1bii (see for example Müller, 2007; Burleson, 2008). An alternative draft had stated simply that mitigation actions by developing countries should be MRV and did not extend this to technology, financing and capacity building from the developed countries. The clause that was adopted, was accepted by developing countries because they felt that it demanded this. In fact, the clause as adopted includes commas that create an ambiguity that could one day cause trouble. As things stand, however, the clause is seen as demanding that developed-country assistance for MRV mitigation should itself be MRV.

Questions remain. Grubb (2008, pp31–32) implies that 'nationally appropriate mitigation … actions' (known as NAMAs) is open to interpretation. This is true. India, for example, has asserted in a submission to the UNFCCC that NAMAs should not include mitigation actions taken without capacity building or other forms of assistance. Grubb adds that mitigation accomplished by carbon trading could introduce a 'distinction between "where to reduce" and "who makes the reduction"', presumably on the basis that emissions reductions in a developing country might, if funded through the Clean Development Mechanism (CDM), be credited to whoever financed them, not to the host country.

The fact remains that the original Framework Convention on Climate Change did not specifically ask that mitigation from developing countries should be measurable, reportable and verifiable, and they had hitherto never agreed to it. At Bali they did. True, they were still not quantifying how *much* abatement they will achieve, and the key demand of the US that they should agree to such concrete targets remains unsatisfied. In theory, a developing country could conform to the letter of 1bii by planting a single tree, provided that action was verifiable and the resulting carbon sequestration measurable and reportable. But, as the South African delegate said at Bali, 'Developing countries are saying voluntarily that we are willing to commit ourselves to measurable, reportable and verifiable mitigation actions. … A year ago, it was totally unthinkable' (Müller, 2007, p5). The developed countries in return agreed to apply the MRV principle to capacity-building assistance that, although promised under the original climate-change convention, had never really materialized.

At the time, the Bali Roadmap was seen as the key that would enable agreement on a replacement to the Kyoto Protocol at COP15 in Copenhagen two years later. As we now know, no such concrete agreement emerged from Copenhagen. Nonetheless, the step forward taken at Bali should not be underestimated.

And another step was taken at Bali – one that may have great importance for agricultural sinks. This was the agreement that any 'Kyoto II' would include credits, not just for reforestation and afforestation as agreed under Kyoto, but *avoided* deforestation. The potential implications of this for agriculture are discussed in the chapter that follows.

Summing up

There were three questions at the beginning of this chapter. First, can we price carbon in monetary terms? The constraint is that economists have based value on the price that will be paid in the market, using the theory of marginal utility. This chapter argued that that theory is itself relatively new, and that economics can change its view as the world changes around it; if we now need to assign monetary values to GHGs, we can. However, we will have to fix the actual prices ourselves, not leave it to God, as St Thomas Aquinas might have suggested. Setting a shadow price is theoretically possible but very difficult, and the figures will be too open to argument to translate into policy. It has been suggested that, instead, the theory of marginal utility needs to evolve so that it applies more clearly to *collective* resource use. In effect, we must do with the atmosphere what pastoralists do with grazing resources, deciding how much we can emit and allocating permits. We can then use the market to set the price.

So we *can* price carbon; but the second question was whether we *should*. There are ethical objections to doing so. Some are far outside the mainstream; it is easy to set up straw men. But others are harder to dismiss. In particular, it will simply not be possible for everyone to one day emit at the same rate as consumers currently do in wealthy countries. So decisions must be made, and this implies the need for valuation. Not everyone who accepts that would also accept that resources should be traded in a market, but that is the inescapable logic of accepting limits to growth.

Finally, this chapter has also discussed the difficult question of whether paying developing countries for mitigation actions is morally justified. It seems that there are pressing practical reasons for taking this route, not least that worthwhile mitigation will not be possible if it does not involve developing countries. As has been seen, in December 2007 developing countries did – more or less – commit themselves to measurable emissions reductions.

The next chapter reviews whether the market instruments for this exist, and how they might apply to agriculture.

4

Flexible Instruments, Fungible Carbon

Since GHGs mix ... in the atmosphere, it is equivalent from an environmental standpoint to reduce emissions anywhere in the world regardless of political jurisdiction. Most of the regulations constraining GHG emissions take advantage of this property of 'substitutability' and allow for the purchase of emission credits both within and outside of the regulated area, thereby laying the ground for a global 'carbon market'. (Lecoq and Capoor, 2005)

In the previous chapter, it was asserted that the market was the best way to reduce emissions. Some readers of this book will, quite reasonably, question this; and there are other models of payment for ecosystem (or environmental) services, or PES, that are applicable to agriculture – they will be discussed later. But this chapter briefly reviews the *market* instruments that have been set up so far, and where agriculture can fit in.

This chapter draws on a number of sources, but the annual report on the carbon market by the World Bank has been especially useful. For several years, the World Bank's Franck Lecoq and Karan Capoor monitored movements in the carbon market, presenting their findings in the late spring in their annual *State and Trends of the Carbon Market*. In 2006 and 2007 the report was prepared by Capoor with Philippe Ambrosi (Capoor and Ambrosi, 2006a, 2007). The most recent (2010) is by Alexandre Kossoy, again with Ambrosi. This report is a very accessible reference for those who are interested in the carbon market, which is already acquiring an insiders' jargon of its own (for example, the year of an emission credit's promised delivery is referred to as its 'vintage').

There are two basic types of carbon-trading instrument: allowance-based, and project-based. As Lecoq and Capoor explain (2005, p11), in the first model, emitters are allocated permits to emit but can trade these between themselves if they so choose. A project-based instrument, however, is issued in recognition of a reduction of emissions. Therefore, 'project-based transactions allow for the creation of new assets that can be used for compliance, above and beyond the initial supply of allowances' (ibid.). (The distinction is a little blurred in that credits generated by projects can be used to augment emissions permits under an allowance scheme, and there are already several examples of this.)

It can be argued that there is a further qualitative difference between two types of project-based allowances: emissions-reduction projects (for example, improvement of generating capacity), which keep some carbon out of the atmosphere, and sinks projects that actually claw it back. A further subdivision might be between sinks that preserve or augment terrestrial carbon in regions where it is relatively stable (boreal forests, perhaps, or temperate forest), and those that do so in fragile regions where sinks could flip over into sources – for example, in the semi-arid steppe. The ecological significance of this was explained in Chapter 2. These additional distinctions appear not to have been much remarked upon, yet there is a powerful argument here for the differentiation of price depending on how a certified emissions reduction (CER) has been created. However, an agreed system for making such distinctions would be very difficult. The main existing market instruments are described below.

The CDM

The Clean Development Mechanism (CDM) is project-based, and is described here first because it is the instrument most often associated with projects in the developing world. Although it is still in operation at the time of writing, its future is not assured; it is part of the Kyoto Protocol, and no firm decision to renew this emerged from COP16 in Cancun in December 2010. However, much has been learned from its design and operation, and this is bound to influence future mechanisms.

Besides reductions of emissions from power stations, industry and other sources, the CDM also covers sinks under the heading land use, land use change and forestry (LULUCF). During the first commitment period, however, LULUCF projects are only eligible if they are those described by article 3.3 of the Kyoto Protocol, mainly reforestation and afforestation. Agricultural sinks are covered by article 3.4 and are therefore not eligible.

This decision was made only in the Marrakech Accords in 2001, and reflected a number of concerns over leakage, permanence, sovereignty, measurements and other methodological concerns (Schlamadinger et al, 2007a, p297). But it also came about because of a general wish to limit credits from LULUCF; this in turn arose from the fact that other aspects of the CDM had already been negotiated, and the introduction of LULUCF late in the stage as an incentive for some negotiating parties rather alarmed others, who felt it might distort an agreement they thought they had already reached.

This suggests a rather ad-hoc, confused way of making international environmental law. But it also suggests there may be no fundamental objection to inclusion of article 3.4 activities in the future. Certainly, in a series of key informant interviews in Washington in 2005, the author found little objection to agricultural sinks in principle. As FAO (2002c, p196) puts it, 'The institutional framework and rules for the global management of climate change are still in considerable flux, and soil carbon sequestration may be eligible for credits under the CDM in future commitment periods.' This has not changed since.

Moreover, there was also growing pressure, especially in Latin America, for the inclusion of avoided deforestation. This too was not included in the CDM at the Marrakesh negotiations, because of concerns about leakage and because the amounts of credits involved could be so large (Schlamadinger et al, 2007b, p278). The latter reason seems curious as it is tantamount to saying that it would involve too much mitigation, but arguably it might have lowered prices for credits and thus prevented abatement in other sectors. This was not the end of the story regarding avoided deforestation, and it does have some implications for agriculture, especially with the development of what is known as REDD+. This is discussed at the end of the next chapter. However, avoided deforestation has never been eligible for the CDM.

The CDM was inspired by Brazilian diplomacy. As Viola (2004, pp38–39) explains, Brazil's policy on climate change accepts it as a major threat requiring urgent action at global level, but it also stresses historic responsibility. Brazilians also have profound sovereignty concerns regarding the Amazonian rainforest, and during the 1980s became increasingly concerned at the way in which this was raised by international partners, such as the World Bank, damaging Brazil's position during bilateral and multilateral loan negotiations (Kolk, 1998, p1485). While in Brazil in

2005 the author saw a poll in a national newspaper suggesting that 71 per cent of Brazilians think the country might be subject to armed foreign intervention because of its natural resources (there is no reason to suppose that the Brazilian government supports this view). There is also a suspicion in Brazil that foreign concern with the rainforest is inspired by a wish to prevent Brazilian agriculture from competing with that of the United States (Matthey et al, 2004, p19).

However, Brazil has shown sophisticated tactics in response to these concerns. As Kolk points out, Brazil recognized that the environment could also be a potential source of power, in which preservation of the rainforest was used as a bargaining chip. In 1989, not long after the visit of a US congressional delegation, the Brazilian foreign ministry had the idea of organizing a major international conference, the Rio Earth Summit, to improve Brazil's image and leverage some 'green' money (Kolk, 1998, p1485).

Rio took place in 1992 and resulted in, amongst other things, the United Nations Framework Convention on Climate Change (UNFCCC). Then in June 1997, Brazil put forward a proposal to address the problem of non-compliance by developed countries with emissions reductions. They would pay fines into a Clean Development Fund (CDF). Viola (2004, p40) explains that this proposal was well received by lower-income countries but not by developed ones. However, in October of the same year Brazil and the United States unexpectedly advocated a revised plan together. This simply proposed a mechanism through which Annex 1 countries could offset their emissions by funding clean development. This was the CDM, and, stripped of its punitive character, it was widely accepted.

The CDM's direct impact on emissions was never going to be decisive. The percentage of a country's emissions that may be offset through it is capped at a low level. It is also very complicated to administer; Kossoy and Ambrosi (2010, p42) report that it is taking longer and longer to get projects from registration to first issuance (607 days in 2009) and the time to get a project to registration has also increased, albeit by less; '[I]t now takes over three years for the average CDM project to make its way through the regulatory process and issue its first Certified Emissions Reductions,' they state (Kossoy and Ambrosi, 2010, p2). Moreover, although land-use projects are permitted in the restricted form of reforestation and afforestation, they have accounted for less than 1 per cent of the volume, not least because credits from such projects cannot be traded on the European Union's Emissions Trading Scheme (EU ETS), discussed later (Kossoy and Ambrosi, 2010, p25, fn). In fact, the CDM's abatement record is dwarfed by that of the EU ETS.

For the post-2012 period, however, any replacement for the CDM should not be dismissed as a possible instrument for agricultural credits. The Bali decision to include avoided deforestation in the post-2012 regime means that such projects would indirectly involve management of soil fertility to mitigate agricultural expansion. Moreover, should the CDM eventually include agricultural sinks, Brazil's involvement may open opportunities for them that would not be available on the same scale anywhere else. La Rovere and Pereira (2005, p3) record that in 1994, just 1.41 per cent of Brazil's GHG emissions (all gases) were from industry, while large-scale hydropower meant that energy generation accounted for 16.61 per cent. By contrast, LULUCF was responsible for 54.93 per cent *excluding* agriculture, which added a further 25.58 per cent (and from CH_4 and N_2O alone; carbon is not included in this figure). This potential was one reason for implementing the field survey described in this book in Brazil rather than elsewhere.

But the real significance of the CDM may be political. Viola (2004, p40) points out that it enabled Brazil to accept the concept of flexible mechanisms. Brazil's status as a very large developing country gives it great influence in climate negotiations, so arguably this is the CDM's greatest achievement. Besides breaking a deadlock, it may also be becoming a bridge between the Kyoto flexible mechanisms and the markets for non-compliant credits established outside the Kyoto system (especially the EU ETS). If this is the case, its true significance may be out of all proportion to the number of emissions credits it creates.

One other Kyoto project-based scheme that should be mentioned here is Joint Implementation (JI), through which countries that do have caps can create (as opposed to trade) certified emissions reductions (CERs). Thus a British investor might fund the refit of a Siberian power station and then sell the CERs so created, or retire them after offsetting them against its own activities. But this does not concern developing countries, except insofar as it helps define the market for CERs. It did seem at one stage that JI might include those transition economies that are in some respects part of the developing world (Robbins, 2004, p25). However, this was effectively prevented when Kazakhstan decided not to ratify Kyoto as an Annex 1 country – not least through pressure from other developing countries, which feared a precedent for non-Annex 1 caps.[1] There *are* non-CDM, project-based schemes that might be important to agriculture in the developing world, but they lie outside Kyoto and are discussed later.

Allowances

Allowance-based transactions are qualitatively different to project-based allowances. However, there is no theoretical reason why project-generated CERs should not be traded against allowances, and this is coming about.

The European Union's Emissions Trading Scheme (EU ETS), launched in 2004, caps about 12,000 large-scale sources of emissions within the EU, covering about 45 per cent of its emissions, and then allows the resulting allowances to be traded (Taiyab, 2006, p32). This acts as an incentive to increase efficiency, either to keep a plant's emissions within the stated caps, or to accumulate surplus allowances that can then be sold to concerns less able to reduce their emissions. The EU ETS had a pilot phase (2005–2007); this was needed, as mistakes are likely when schemes like the EU ETS first take shape. Indeed, there were bitter protests from the World Wide Fund for Nature and the two main British opposition parties in May 2006 when it turned out that the more efficient power generators had effectively received a huge windfall through the allocation of permits, which, it was said, should have been auctioned, not given away; ironically, this was reported not long after a crash in carbon prices (Harrabin, 2006, p2). These two events were linked in a recent report by an anti-EU think tank in Britain, which suggested that the over-allocation of permits had so weakened prices that there had been no climate mitigation at all (Open Europe, 2007, p5). In fact, Capoor and Ambrosi (2007), while agreeing that the allowances for 2005–2007 were inappropriate, also blame information leaks that some EU countries had more carbon in hand than expected, wiping out over half the EU ETS market value (Capoor and Ambrosi, 2006a, pii; 2007, pp3–4, 12).

Since then, however, trading on the EU ETS has shifted beyond Phase I, and the fact that Phase I credits are now worth little is less relevant. In 2009, $119 billion-worth of allowances and derivates were traded on the EU ETS (Kossoy and Ambrosi, 2010, p2). Moreover, there is some evidence that it has had a real impact on the EU's emissions, perhaps as much as 2–5 per cent during the pilot phase (op. cit., p6). By 2009 there was also substantial US participation in the market for allowances (Kossoy and Ambrosi, 2010, p2), suggesting untapped potential for an allowances scheme in the US itself.

Moreover, the EU ETS does have regulatory power. As from 1 January 2005, all plants involved in certain activities (including energy, iron and steel, minerals and paper) and emitting the GHGs associated with them, must 'be in possession of an appropriate permit issued by the competent authorities'. The European Commission had powers to reject the initial national allocation plans of member states, and did partially do so in five cases (European Commission, 2006, p1). There are monitoring requirements, and the plants are obliged to surrender annually a quantity of allowances equivalent to their emissions. They also face non-compliance fees that will eventually rise to €100 per tonne (Taiyab, 2006, p26).

Perhaps most importantly, CERs garnered through the Kyoto flexible mechanisms – JI and the CDM – can be redeemed against emissions allowances under the EU ETS. The European Union's Directive 2004/101/EC[2] permits plants to offset their emissions using CERs created under these projects to meet their obligations under the EU ETS, helping to meet their countries' Kyoto emissions obligations at the same time. This could massively widen the market for CDM-generated CERs. It cannot increase their supply, beyond a certain point, as the CDM is capped. But it will increase their value.

The linking directive has some important caveats. One is that CERs cannot be obtained from nuclear power generation for this purpose; France's objections to this clause delayed the directive, but in fact such CERs would not be Kyoto-compliant. More to the point, the directive states LULUCF projects cannot be used to generate credits for the EU ETS – although there has been some pressure for this to change (see for example Taiyab, 2006, p13). Although the EU has historically distrusted sinks, the LULUCF exclusion appears to be less ideological than practical; the Kyoto method of accounting for sinks is through temporary credits (T-CERs) to allow for non-permanence, a subject discussed further in the next chapter. This makes it hard to reconcile LULUCF-derived credits with CERs for the EU ETS. In the long run, this may not prevent the linking directive from improving the prospects of agricultural sinks, as credits from the EU ETS and the flexible mechanisms may find themselves being traded in completely different markets, whether the European Commission thinks they should be or not.

How this might theoretically occur may be seen from the New Zealand example. In 2009, New Zealand introduced the only nationwide mandatory scheme outside the EU ETS. From 2015 this will include agriculture, which accounts for nearly half the country's emissions (Kossoy and Ambrosi, 2010, pp23–26). The scheme also permits the import of Kyoto-compliant CERs (Kossoy and Ambrosi, 2010, p25, fn), so there might be potential for New Zealand emitters to buy Kyoto REDD (reduced emissions from deforestation and degradation) credits, with potential gains for agriculture. There are also plans to explore linkages with the EU ETS. By such roundabout routes agricultural carbon credits may become fungible and end up in the EU ETS, the biggest scheme of all.

The New Zealand scheme remains a world leader for now. However, there has also been progress with domestic cap-and-trade schemes in a number of other countries. There are also now emissions trading schemes in New South Wales, the United Kingdom – and in Japan, where Kyoto-based credits can be used against domestic allowances; however, outside Tokyo, the Japanese scheme remains voluntary (Kossoy and Ambrosi, 2010, p28). China has three voluntary exchanges, and India, Mexico and the Republic of Korea also have embryonic carbon markets (Kossoy and Ambrosi, 2010, pp32–34). Perhaps most interesting of all, Brazil is considering a domestic cap-and-trade scheme that will include agriculture that may include restoration of grazing land (Kossoy and Ambrosi, 2010, p32, fn) – an extremely important method of carbon sequestration that is discussed in detail, in the Brazilian context, in Chapters 6–8.

However, a major factor in the carbon market is the participation or otherwise of the United States. It declined to ratify the Kyoto Protocol, but remains a signatory to the UNFCCC itself and is fully involved in the negotiations for a 'Kyoto II'. In the meantime, the election of 2008 raised the possibility of a domestic, nationwide, mandatory cap-and-trade system on the same principle as the New Zealand scheme or EU ETS. This took the form of the Waxman-Markey Bill, which aimed to reduce emissions by 17 per cent below 2005 levels by 2020. It would have permitted quite a high level of offsets – up to 40 per cent of the average annual cap – split equally between domestic and international offsets, including REDD (Kossoy and Ambrosi, 2010, pp29–30). The eligible offsets were never clearly defined, but the potential for agriculture, including foreign agriculture, could have been considerable. The bill did narrowly pass in the House of Representatives, despite considerable opposition (including that of liberal representatives who thought it too weak). However, it failed to navigate the Senate, and a compromise bill proposed instead by Senators Kerry and Lieberman ran out of time with the onset of mid-term elections in November 2010. Given the results of those elections, a mandatory national cap-and-trade system seems unlikely.

Despite the lack of a national scheme, however, the US has in fact made much progress in developing a carbon market. An important early mechanism was the Chicago Climate Exchange, or CCX. Although this was voluntary, transactions made on it were legally binding. The CCX allowed trade in offsets generated in the United States or Brazil, and a large percentage of its business concerned agricultural sequestration projects that would not be welcome elsewhere, albeit not agricultural credits from Brazil (Taiyab, 2006, p34); moreover it tried to encourage cooperation with the EU ETS.

The CCX unfortunately fell victim to the recession of 2008 and the accompanying decline in the carbon market, as well as doubts as to the eligibility of its credits under the proposed national cap-and-trade legislation, and trading of carbon credits ceased at the end of 2010. However, at one stage it had looked as if emissions reductions (ERs)[3] traded on the CCX might one day be worth much more due to its proximity to a growing American market in the shape of states who were unhappy with the US's rejection of Kyoto and were trying, to a greater or lesser extent, to comply, reducing emissions on a regional basis.

This trend has attracted little attention outside the US, the federal nature of which is not always understood abroad. But it is significant. For example, an April 2003 initiative by New York State has now grown into the Regional Greenhouse Gas Initiative (RGGI), under which a large area of the US has a mandatory cap-and-trade system for emissions from power generation, not dissimilar to the EU ETS. Ten north-eastern and mid-Atlantic states are involved. Carbon offsets from other, strictly defined, sources may in principle be added to a power generator's budget, but only up to a very limited percentage of its emissions budget, and agricultural sinks are not eligible. However, they do include emissions of methane from manure, which could in theory be an incentive to move from livestock to more carbon-intensive farming systems. (In practice, there might need to be other incentives as well, such as availability of labour. This type of issue is discussed with reference to Brazil in Chapters 8 and 9.)

In August 2006, the RGGI was joined by another important new law in the State of California requiring the California Air Resources Board

(CARB) to develop a regulatory structure for 'market-based' incentives, with the object of reducing emissions by 25 per cent, bringing them back to 1990 levels, by 2020, and 80 per cent below 1990 levels by 2050; mandatory caps will begin in 2012 (Capoor and Ambrosi, 2006b, p17). The cap-and-trade regulation was expected to come into force by the beginning of 2011 and is likely to permit offsets, including foreign offsets, in principle, but they will be strictly supervised; for the farming sector, they are likely to be REDD and manure management only. However, these do have indirect potential for agricultural sinks.

Two further important regional cap-and-trade mechanisms are also coming on stream. The Western Climate Initiative (WCI) involves a number of states in the western US and four Canadian provinces, and will become operational in January 2012. The Midwestern Greenhouse Gas Reduction Accord (MGGRA), which involves several Midwestern states and also has Canadian participation, is further behind, having agreed to back a federal solution, but with the intention of instituting a cap-and-trade system of its own should federal legislation fail. This is now scheduled for 2012. There are also a number of initiatives taken by individual states. However, these schemes are not yet completely certain. For example, two states have recently defected from the WCI, and the others face legislative hurdles to their participation; states in the MGGRA, too, are at different legislative stages (Hamilton et al, 2010, p14).

Fungibility

Arguably, the significance of the US developments is threefold. First, the RGGI and other regional initiatives will ultimately create a large market for carbon offsets – including, to some extent, foreign ones. Second, both agriculture and developing countries are on the agenda. Third, there is a willingness in principle to trade in Kyoto-compliant offsets, and this will presumably extend to whatever replaces the Protocol, if anything does. This cooperation may never be reciprocated, but that may not matter; a buyer will be quite willing to give up a CER in return for an ER if s/he has equal confidence in the latter. This is no different to my being happy to accept Euros or dollars in a transaction, even though they are not legal tender in Britain. True, my own central bank did not issue them; but I know that it has confidence in whoever did, and will exchange them. Carbon offsets, then, may become a fungible commodity, raising demand and value.

But there will be a price to pay for fungibility. A buyer may hold (say) 2000 tonnes of CO_2 offsets but will have no control over, and perhaps no idea, of how they have been generated, so s/he will not know whether sustainable development criteria demanded by the CDM has been met.

It could be argued that this does not matter, as all emissions mix together in the atmosphere in the end. However, while the reforestation of degraded pasture could have been with tropical hardwoods, it could also have been done with fast-growing eucalypts, which would generate credits far more quickly but will have a high water requirement and will add little biodiversity. Even the impact on climate change may be variable; it was stated earlier that agricultural sequestration projects in fragile semi-arid regions were more valuable in terms of climate change than those in temperate zones, as they might prevent a sink from becoming a source. In a world of fungible carbon, that distinction will be lost. It also means that there will be cheaper ways of generating offsets than agriculture, especially small-scale agriculture in low-income countries. How this can be dealt with is for the next chapter, which examines agriculture in the carbon market.

However, before leaving the subject of the market itself, it is worth mentioning voluntary offsets.

Voluntary Offsets

Voluntary offsets generate emissions reductions (ERs) rather than CERs; these are less fungible, as there is no mandatory market on which they can be sold to emitters to offset their legal requirements. These offsets drive the retail carbon market – something completely outside the mechanisms that have been described so far. This consists of individuals and small companies that are not subject to any form of statutory cap, but who wish to offset their activities for ethical or public-relations reasons. For example, a family concerned about climate change can offset (say) the emissions from their holiday flight – this is a common use – by buying offsets from a retailer, often online.

The voluntary sector is tiny compared to giants such as the EU ETS. Hamilton et al (2010, pii) put it at 93.7Mt of CO_2e in 2009. Kossoy and Ambrosi (2010, p1) put it lower, at just 46Mt of CO_2e in 2009, worth $338 million – compared with 8700Mt and just under $144,000 million for the entire global carbon market. Both sources agree that, unlike the market as a whole, it declined against 2008 – suggesting that it is more recession-sensitive.

ERs from the voluntary market will not normally be Kyoto-compliant. Lecoq and Capoor (2005, p13) say they may have been generated in accordance with CDM procedures, but in view of the CDM's complexity this seems unlikely. This raises some quality questions. Duncan MacLaren, writing in *The Scotsman* (7 December 2004), points out that forestry schemes can be especially doubtful. 'Some firms get payments just to avoid felling existing forest. Many other new trees would have been planted anyway. Some are planted on carbon-rich peat bogs – emitting more carbon than they sink,' he states, and goes on to claim Friends of the Earth International have found projects where primary rainforest has been felled and replaced by plantations.

There is thus an ethical dimension lacking in some consumer protection, in that offsets generated in poorer countries will have livelihoods impacts that the buyer does not see. Partly for this reason, some retailers avoid LULUCF projects or restrict them to 20–25 per cent of their portfolio. Where sinks are used, a questionable practice is the sale, in some cases, of 100 years' worth of carbon from a forest in advance. What is to protect such an investment? As this permanence question is especially relevant to agriculture, it is discussed in the next chapter. A further cause for concern is the very large mark-up, with some retailers spending only 25–30 per cent on the project and as much again on marketing. Capoor and Ambrosi (2006b, p10) say that the 'single biggest impediment to stronger demand … remained the lack of a broadly accepted standard for voluntary projects that combined simplicity and consistent integrity'.

This has since been partially addressed by the rapid growth of third-party certification bodies. These cannot make activities Kyoto-compliant if, like agriculture, they are not of a type accepted under the Protocol. But they can provide an independent verification of a genuine emission reduction. They might even be able to provide pre-certification for compliance with national standards currently under development (although that would be highly uncertain).

The best-known third-party body is the Gold Standard, which has participation from the private sector as well as NGOs. This not only demands involvement of a certification body for CDM/JI but also sets sustainable development criteria that the CDM Executive Board does not require. However, the Gold Standard also imposes the same exclusions as would exist on an 'official' trade – it does not permit LULUCF projects, for example (Butzengeiger, 2005, p16; Gold Standard, 2009, p29). This seems illogical when it is LULUCF projects that most need such screening.

The Voluntary Carbon Standard (VCS) *does* include agriculture. Although founded only in 2007, it has now overtaken the Gold Standard in popularity as a certification body for the voluntary market, accounting for 35 per cent of retail projects (Hamilton et al, 2010, pviii). It was founded by a coalition of three international non-profit and sustainable development organizations that included the International Emissions Trading Association (IETA).

The VCS includes agricultural activities under the banner of Agriculture, Forestry and Land Use (AFOLU). VCS guidance for AFOLU specifically includes cropland management with 'practices that demonstrably reduce net GHG emissions from a defined land area by increasing soil carbon stocks, reducing soil N_2O emissions, and/or reducing CH_4 emissions'. It also includes improved grassland management, 'including the adoption of practices that increase soil carbon stocks and/or reduce N_2O and CH_4 emissions' (VCS, 2008a, pp2–3).

The VCS requirements must be met for certification; it has, for example, a 'buffer' percentage of credits that are withheld with certain types of AFOLU activities, against the risk of non-permanence. There is of course also a baseline requirement, but it is flexible; a measured soil carbon content can be used if appropriate data is available, but so can activity-based measurement (VCS, 2008b, p21). This would be helpful in dealing with some of the issues that will be dealt with in the next chapter.

Hamilton et al (2010, pp57–67) quote over 20 third-party standards bodies for the voluntary market, including the VCS and the Gold Standard. Some are associated with particular carbon markets or carbon types; they do not necessarily certify agricultural sinks (although one or two do certify projects that reduce methane from agriculture). The requirements of all these standards are subject to change, and there is no point in listing them here. But several are worthy of mention.

One is the Panda Standard, a collaboration between Chinese and international organizations – the latter include the well-known NGO Winrock International, which has also pioneered the American Carbon Standard. Announced at the Copenhagen summit in 2009, the Panda Standard is designed for the potentially enormous Chinese market. It requires both that projects be MRV, and that they provide ancillary benefits. AFOLU projects are eligible, but there is also a permanence requirement, which has difficult implications for agricultural sinks; again, this is discussed in the next chapter. However, the Panda Standard appears to allow for this, as it includes a buffer pool mechanism similar in principle to that of the VCS (Panda Standard, 2009, p9). Moreover, the Panda Standard released a set of sectoral specifications for AFOLU in January 2011.

What is especially attractive about the Panda Standard is its firm Chinese base and apparent host-government support. This would suggest that, in the event of China instituting a nationwide cap-and-trade system in the future, ERs in the Panda system might be bought for pre-compliance; such credits would have a good market value. That is speculation for now; and besides, the standard is designed for China and will not certify projects anywhere else. Nonetheless it is an interesting development.

Other certification bodies that are worth mentioning are those certifying on the 'stacking' principle; that is to say, they certify not actual carbon mitigation, but the social and environmental co-benefits of the project. Thus a project might be certified for mitigation purposes by (say) VCS; a body such as SOCIALCARBON will then vouch for its other aspects. SOCIALCARBON is Brazilian but may expand outside Brazil in the future (Hamilton et al, 2010, p64). Other schemes include the CCB Standards, run by the Climate, Community and Biodiversity Alliance, and supported by several international NGOs. These 'stacking' mechanisms blur the distinction between the carbon market, on the one hand, and non-market payments for environmental services on the other – these are discussed in the next chapter.

The retail carbon sector will grow, and demands for regulation will grow with it; and, in time, non-compliant retail ERs are bound to find their way into compliance markets or to be counted in the next phase of the RGGI. Interestingly, Hamilton et al (2010, pxi) report that 'A significant chunk of credits transacted in 2009 went to pre-compliance buyers focused on buying credits that might be eligible in a future compliance market.' As these markets grow, it will eventually become worthwhile for EU ETS allowances to be retired against the ERs traded on them.

Moreover, the share of the voluntary sector taken by agriculture is higher – and growing; Hamilton et al (2010, ppvi, 31) put agricultural soils at 3 per cent, or 1.2Mt CO_2e, up from just 0.5 per cent in 2008.

But this is still quite low; also, they report an average price for agricultural soil projects of \$1.20/t CO_2e, low against most other types of project but especially solar energy (\$33.80/t CO_2e). A tonne of carbon might be just that when it is in the atmosphere, but mitigating it has different costs depending on the method used. Hamilton et al also point out that most of the 'voluntary' soil carbon has so far been transacted through the CCX, and although this is changing, they say they have yet to hear of a soil carbon project being certified outside it (Hamilton et al, 2010, p35). This makes it yet more urgent that tools are agreed for the verification and monitoring of soil carbon against agreed baselines. This will be discussed in the next chapter.

The carbon market is in its infancy, and has suffered a setback because of the 2008 recession. But it will grow and, importantly, Kyoto-compliant CERs look likely to be only part of it. This means there will be space for ERs generated through non-compliant LULUCF or AFOLU, such as agriculture; meanwhile the eligibility of REDD means that Kyoto II-compliant mechanisms might now include agriculture to a degree.

But the market for project-generated carbon will in the end be far greater than that generated by Kyoto alone. The failure of the US to ratify Kyoto was nearly a disaster, because it could have meant it did not come into force. Given that Kyoto did eventually do so, however, the absence of the US from it will in the end have helped a far larger, flexible and more vibrant carbon market than would otherwise have been the case. The keys to its success will be the quality in terms of reliable emissions reductions, transparency and honesty.

Summing Up: Into the Market

Having established in the previous chapter that carbon first could, and second should, be priced and traded, in this chapter the types of market instrument were reviewed.

The creation of this market has been messy, and it remains uncertain and fragmented. What is important for agriculture, however, is that the US did not ratify Kyoto. Had it done so, agricultural sinks, as a non-compliant mitigation strategy, might have had no market in the foreseeable future, as the Kyoto flexible mechanisms would have been greatly strengthened – and they exclude agriculture. The fact that the US is now creating its own markets means that such sinks soon may have a market, and indeed may become indirectly eligible for CERs issued under Kyoto or its successors, if any, through the growing fungibility of carbon credits as they are traded between different mechanisms.

Indeed, given the lengthy approval processes involved in the CDM and Kyoto's cautious approach to sinks, the volume of GHGs traded worldwide may soon be much higher than it would had the US ratified, as the Kyoto mechanisms would then have dominated the market, and, through their bureaucracy, limited it. This is not meant to imply that the refusal to ratify was in any way helpful. Had it destroyed Kyoto, as it nearly did, the global carbon market would have gone back to square one. As things have worked out, however, US withdrawal has not been an entirely bad thing.

This chapter has argued a market is emerging. The next chapter will discuss how agriculture in the developing world can enter that market, and will look at issues such as baselines, leakage, permanence and measurement. It will also discuss non-market mechanisms that provide payment for ecosystem (or environmental) services (PES) – a potentially important alternative.

Notes

1 In fact, in 2010 Kazakhstan did submit a proposal to the UNFCCC according to which it would accept emissions caps and accede to Annex B of the Kyoto Protocol.
2 Or, to give it its correct title, 'Directive 2004/101/EC of the European Parliament and of the Council of 27 October 2004 amending Directive 2003/87/EC'.
3 There is an 'alphabet soup' around the naming of the different types of unit. Lecoq and Capoor (2005) use the acronym ER for a non-Kyoto-compliant emissions reduction, as distinct from a certified emissions reduction (CER), which would count against a country's Kyoto or ETS commitments, or a verified emissions reduction (VER), which the buyer hopes will become a CER. Butzengeiger (2005) and Taiyab (2006), however, use VER to indicate a non-compliant unit. The author has followed Lecoq and Capoor.

5

Carbon, Money and Agriculture

Scientists are very interested in participating in the development of soil C monitoring protocols. However, in order to contribute to the development of these monitoring mechanisms, they need to know the economic and policy rules under which these mechanisms are anticipated to operate. Because so much is at stake, a multi-sector, multi-discipline, and multi-national effort is required if we are to make monitoring and verification of C sequestration in soils a useful and widely used procedure. (Post et al, 2001)

Can agriculture in the developing world participate in the emerging carbon market? If so, how? There is very little experience with agricultural carbon projects, and not much more with LULUCF, although the latter can already be seen to raise equity issues (Boyd, 2002; Brown and Corbera, 2003; Brown et al, 2004; May et al, 2004); agricultural sinks might also have livelihoods implications, but these are reviewed later in this book.

There *is* a very large literature on the agronomic techniques that would be needed to sequester/retain carbon, but usually not written in a carbon context. This chapter looks at how carbon projects could be constructed in agriculture: how the carbon could be traded, project design for permanence and baselines, and monitoring.

There are three main subject areas for this chapter. The first is whether agricultural sinks can be monetized under the existing market mechanisms. The second is what issues would have to be solved in the project design.

The third area looks at less rigid types of project, perhaps publicly funded, that can be used if the methodological issues involved in producing CERs are too difficult. There are already a number of payment for environmental, or ecosystem, services (PES) initiatives under way, and if the monitoring and verification demands of the formal allowance market are too difficult, that could be an alternative. This chapter is structured to consider the following three questions:

- Would anyone invest in agricultural sinks in the developing world, and through what market mechanisms?
- Could the projects overcome some of the problems especially associated with LULUCF projects – baselines, additionality, leakage, measurement and monitoring?
- Can PES schemes offer farmers an effective alternative to a market-driven system?

Who Would Invest?

The constraints to investment in agricultural sinks are partly those of project-based credits themselves. As Lecoq and Capoor (2005, p11) explain, a CER generated by a project is no different from an allowance once it is issued; however, formal certification depends on project performance and there is an inherent risk. The extent of that risk, and the price per tonne, matters more to buyers than the details of how the carbon credits were generated.

This was apparent at the inception of the market. The Kyoto Protocol came into operation in 2005, and although 43 million tonnes worth of projects were transacted in the first four months of that year (Lecoq and Capoor, 2005, p19), most buyers made it clear that they were buying CERs, not investing in the project that created them. The project developer was expected to bear the risk of the emissions reduction not ultimately being registered as a CER, and therefore not having 'official' value.

Where the buyer does agree to bear the risk of a VER not becoming a CER, the price per tonne for VERs is lower, and in some cases – where the technology is difficult, or there is a serious leakage risk, for example – it might be much lower. Some contracts have apparently had provisions under which the buyer would be compensated in cash at the prevailing spot price for any shortfall in tonnes registered (Capoor and Ambrosi, 2006b, p10) – an interesting gamble on both sides.

A further constraint is uncertain delivery dates. If an investor purchases VERs on the understanding that they will be certified as CERs on a given date, they take a risk of non-certification because of delays in project implementation. This means it may not be possible to use them for compliance in the way that was originally intended. In fact, the fungibility of credits and the gradual convergence of climate regimes, discussed in the previous section, means they will probably be able to cover their needs by 'buying-in' secondary CERs and sell their own later, upon certification, on the same market; but that is not assured. And the risk of non-certification will remain in any system that trades in statutory allowances; so will the need for ERs to be delivered in the year for which they were contracted.

There are special problems with LULUCF in general, and agriculture in particular. The period to certification would be even more of a constraint; it takes a long time to build up SOM. Moreover there would be even more difficulty attracting investment up-front. Abatement costs per tonne vary enormously; agriculture would not be the cheapest method, and the poorer countries would not be the cheapest places to do it. Thus early projects were often for destruction of HFC_{23} (a powerful GHG which is a by-product of refrigerant and Teflon production) as 'low-hanging fruit', and easier types of emissions reductions will always be contracted for before difficult projects such as land use. This became clear as soon as the CDM was launched; Capoor and Ambrosi (2006a, 7) report that 58 per cent of the volumes in project-based transactions between January 2005 and March 2006 were for the destruction of HFC_{23}. Land-fill gas was next at 18 per cent, then coal mine methane (6 per cent); N_2O, hydro, wind, and biomass all accounted for 3 per cent each. Agroforestry and LULUCF together accounted for just 1 per cent. This was of course in part through agriculture's ineligibility for the CDM. But some forms of LULUCF were already eligible, and did not prove attractive to investors.

A further difficulty is that there is a large theoretical problem in calculating the real mitigation achieved by sinks projects. This is because there is an essential difference from mitigation achieved by a simple emissions reduction; in the latter case, a tonne of carbon was going to enter the atmosphere, but was prevented, and now won't do so. In the case of a sinks project, however, the emission may simply have been delayed. On the other hand the reduction *might* turn out to be permanent. This presents a problem in accounting for the value of sinks which is discussed later in this chapter.

A further question that must be considered with agriculture is that abatement costs are also likely to be highest in cases of greatest need. It will be more expensive to build up carbon in highly degraded soils, for example. 'Thus, areas of land that may have the greatest physical potential to supply soil carbon sequestration may also be those where it is most expensive,' says FAO (2002b, p203). Also, arrangements with low-income farmers who could most benefit from carbon income are likely to have the highest transaction costs, which are 'obviously much higher when dealing with small and geographically scattered producers operating under heterogeneous agro-ecological and institutional conditions ... [This] is a key issue that must be addressed in order to channel the benefits of such programmes to the poor' (op. cit., p207). FAO suggests bringing farmers together in trading blocs based around pre-existing organizations such as farmers' associations, local governments and NGOs. But it also sees problems in the heterogeneous property rights of low-income farmers.

The answer to all this is likely to be up-front financing from someone other than the eventual purchaser of the CERs. There are two ways in which this might be done – through a unilateral project, or through a development project that might subsequently sell the carbon. There is clearly some crossover between the two.

Before the CDM was even launched, it was clear that different types of project might need different funding mechanisms. Baumert et al (2000, pp1–3) define the different project designs as bilateral, multilateral and unilateral, the differences between the three being defined in part by the relationship between the project's design and financing, and whoever eventually uses the CERs.

Bilateral CDMs, they explain, would involve the Annex 1 investors participating directly in project design and financing, and possibly implementation. On this model, however, they predicted that investment would be likely to follow more general foreign direct investment (FDI) in the developing world, with over 50 per cent split between Brazil, China and India (Baumert et al, 2000, p4), and for similar reasons (existence of mechanisms, transaction costs). This prediction quickly proved correct. Moreover the average transaction volumes greatly exceeded what most agricultural projects would achieve, unless on a wide regional basis. If the CDM or similar mechanism eventually admits agriculture, how can this be overcome?

Baumert et al (2000, pp4–5) consider the question of 'poor value' projects. Their discussion centred on the CDM, which was then in the future but may now nearly have run its course in its present form. However, their reasoning would hold true for any project-based marketplace, so is still worth repeating.

They discuss two alternative models to the bilateral CDM. One is the multilateral CDM, in which investors would pay into a fund that managed a portfolio of projects administered according to principles agreed between the parties. This would pay out proportionally. Thus projects with lower returns and higher costs could be 'buried' as part of the whole. This would also spread the risk to investors of non-certification and other problems. Some portfolios did emerge quite early in the CDM (Capoor and Ambrosi, 2006a, p29), so some commercial investors may choose to include 'ethical' types of mitigation project in the mix, either to improve their corporate image or to be able to charge investors a premium for what might be called fair-trade carbon. (Or, as it has more recently been called, 'gourmet carbon'!)

However, it is the other alternative, the unilateral CDM (or other mechanism), that looks more promising for less efficient carbon suppliers such as African farmers. Here the host country simply mounts the project themselves and sells the credits afterwards, so the end users of the credits do not bear the higher mitigation costs, and would not care about them. Given that so many investors already seem to be trading secondary CERs without caring too much where they have come from, this suits the needs of both sides quite well.

Unilateral projects do not give the project developer complete freedom; it still needs to get its project design approved for the CDM or whatever replaces it. Other mechanisms such as the RGGI would make similar demands, though they might not be so stringent; reputable voluntary carbon standards such as the VCS would also do so. The point of the unilateral project for the developer, however, is that they can eschew market-based investment and mount the project from national development funds, or with loans from an agricultural bank. This helps not only those countries that are uncompetitive through transaction and mitigation costs, but also those that want to sell carbon sinks that are high risk for other reasons, such as security – Baumert et al use Colombia as an example (Baumert et al, 2000, p7). Jahn et al (2003, p6) point out that in such cases, the host country might be better able to assess such risks and may implement the project whereas an outside investor would not.

The unilateral model was accepted by the Kyoto parties at COP7 in Marrakech in 2001. Its advantages are such that it has been implemented not just by the poorest countries, but those that were expected to do well out of the CDM anyway. It proved its worth as a model straight away; of 107 CDM projects approved by India's national authority by November 2005, 91 were unilateral, including three of the seven approved by the

CDM Executive Board (EB), and there were more in the pipeline (Jung et al, 2005, p24).

Some Indian commentators have expressed concern that this is not really in the spirit of the CDM, which was meant to involve development and technology transfer (Jung et al, 2005, p24). Jahn et al (2003, p9) identify this as a disadvantage of unilateral CDMs and suggest others, including many that would affect any unilateral mitigation project. For example, the project developer risks a decline in the price of CERs/ERs between project inception and bringing them to market (although the opposite could also apply). Local investors may also find it harder to obtain insurance coverage against project failure, as they may not have the financial infrastructure that would be available in Annex 1 countries (Jahn et al, 2003, p11). A local developer may also not have the expertise in (for example) setting baselines.

However, the major challenge for unilateral ER producers will be simply financing the project at its inception. Official development assistance (ODA) could be used for capacity building in project design, but cannot be used to finance CDM projects themselves, and will almost certainly not be allowed to under any successor mechanism. But although ODA cannot be used to finance CDM projects, there is a grey area where the CERs from ODA might be accepted if they accrue to the host country and not the investor. This is not conceptually different from giving development aid to an industrial sector – after all, the donor does not get the products.

More likely, however, land-management or other projects might be financed for national development purposes, and any CERs or ERs considered as by-products to be sold at the end of the project. In any case, under a universal cap-and-trade regime, any increase in sectoral carbon sequestration would count against a country's emissions and could be sold as emissions allowances, even if they were not used to generate project-based emissions reductions. It should be noted, however, that there is a danger of double-counting, if they were to be considered as both.

Suitable sources of such funding might include World Bank co-financing, especially with incremental funding from the Global Environmental Facility (GEF). GEF does not finance projects as such but provides incremental funding for existing proposals in order to increase global environmental benefits. Thus a sectoral food-production programme, funded to assist agriculture and reduce malnutrition, might use such extra inputs from GEF to incorporate biodiversity or carbon-sequestration concerns. An example of this was the Kazakhstan Drylands Management Project, which was expected to generate $5.7 million worth of carbon; there were no firm plans to sell this, but it was considered a possibility.

But there are disadvantages to this sort of funding. Organizations like GEF have to work through formal structures at high level, and this makes them less responsive than the market to the needs of farmers and investors. A senior official of GEF told me in 2005 that there would be considerable problems with market-based mechanisms, but thought they were a better alternative to GEF-type funding where possible. 'The GEF is less efficient in this sense – we are geared to deal with governments and our projects take five or six years to become a reality,' he said. 'Farmers can't deal with that. International assistance moves to different rhythms. But a farmer – once he makes a decision, he acts.'

This should not be taken as a criticism of GEF; its purpose is different from a market mechanism. This actually confers an important advantage. A market/compliance mechanism must have an account of exactly what has been delivered in tonnes per hectare of carbon. GEF is more flexible and can bundle carbon sequestration with other environmental goals such as biodiversity and watershed protection. It must still cost the global benefits of these, but need not do so quite so accurately, making the integration of the different benefits easier. This not only adds value to a project; it also addresses the problem of untangling one type of environmental benefit from another for accounting purposes, a serious constraint to any intervention purporting to provide a specific environmental service. This is one advantage of the non-market PES model, which will be described later in this chapter.

Before that, however, it is important to look at the demands that any investor, public or private, and any certification mechanism, will make on farmers. As explained at the end of Chapter 3, a crucial agreement was reached between North and South at Bali: that mitigation actions must be 'measurable, reportable and verifiable' (MRV). Some of the technical demands discussed below will therefore be made on a wide range of projects, not just those that generate CERs.

Additionality and Baselines

The M in MRV – measurable – implies several things. First, one must be able to measure the net emissions reduction or GHG removal. To do this, two things must be clear: first, the rate of emission, or size of the sink, at project inception (the baseline); and second, whether any subsequent changes are the result of the project – or would have happened anyway (additionality). Both present serious problems for land-use projects of any kind, and for agriculture they will be especially complex.

For the additionality requirement to be satisfied, it must be demonstrated that emissions are lower, or sink storage higher, than they would have been under 'business-as-usual' (BAU). This does not *necessarily* mean that emissions should be lower at the end than at the start of the project period (in the case of Kyoto, baselines run from 1990). If there is an increase in emissions, but it is lower than that projected in the baseline, then there is still additionality.

In the case of (say) a power station, one might calculate the baseline by using consumption projections that would be compiled anyway to plan plant provision. In the case of agriculture, the picture is more complicated, as carbon-friendly practices have often been devised for other purposes, and it is hard to prove that they would not have occurred under BAU. Noble and Scholes (2001, p17) use a minimum tillage project as an example of this difficulty. 'Firstly, accurate estimates of the emission reduction in 1990 are needed and these may be difficult to determine now,' they say. 'Also some feel that minimum tillage was introduced for reasons other than mitigating climate change and that under a business-as-usual scenario would have increased in application. They would prefer to credit only emission reductions above such a scenario.'

There is ample reason to support this statement. Reduced, or conservation, tillage was mentioned briefly in Chapter 2, and is discussed in more detail in Chapter 8. In essence, however, it is the sowing of crops with no or minimum ploughing. It has had some support in recent years as a technique that can increase soil carbon content, but it was not originally developed for that, but for soil conservation. In southern states of Brazil such as Paraná and Rio Grande do Sul, farmers turned to no-till and reduced tillage in the 1970s to repair the damage of years of intensive farming; as FAO (2001b, p1) puts it, 'nature presented its bill'. Boddey et al (2003, p605) report a high level of enthusiasm for the technique: 'Brazilian farmers have … taken to no-till with such enthusiasm that they have formed associations with names such as "*Clube da Minhoca*" (the Earthworm Club) and "*Amigos da Terra*" (Friends of the Land), and now pay close attention to sustainability, to the extent that they plant winter crops solely to improve the subsequent maize or soybean crop.'

This makes it hard to claim additionality, and the same might apply with agricultural intensification if it could be argued that it would have been implemented anyway for food security. As Ringius (1999, p27) says in his discussion of CDM SOC potential in Africa, a narrow interpretation of additionality would reject any revenue-generating project because it would have happened anyway.

But this would exclude most projects that increased sustainability through more rational use of the natural-resource base. An analysis by Pretty et al (2006) seems to demonstrate this, examining 198 projects centred mainly on water-use efficiency, organic matter/carbon sequestration and reduced use of inputs to control pests and diseases. They found that 79 per cent showed yield increases across a wide range of crops and systems.

Sugiyama and Michaelowa (2001, pp76–77) refer to something similar, called 'Grubb's Paradox' after the book in which this was raised (Grubb et al, 1998). Crudely stated, the paradox is that the better a project is in terms of co-benefits (which are required under the CDM), the harder it will be to prove additionality. In the case of the CDM, Sugiyama and Michaelowa used Grubb's Paradox to demonstrate how the project with the greatest additionality would have the highest mitigation costs per tonne of carbon. This is especially illogical with agriculture projects, as it implies additionality can only come from using practices that farmers would explicitly *not* have adopted without carbon funding. If induced to do so solely through incentives, they will drop them after the project, and the carbon sink will be short-lived. (This is discussed further in later chapters of this book.)

However, as Ringius points out, there can be barriers to projects (funding, expertise, institutional constraints) that the CDM can help overcome. Without this intervention, perhaps the project would *not* happen, even where the practices are attractive to farmers, so there is additionality. Ringius suggests a pragmatic case-by-case assessment, and this is the only sensible solution.

Where projects are funded through GEF, the additionality problem will still exist in another form, as global benefits (such as carbon) must be expected from a project before incremental funding can be awarded. GEF (2005a) explain: 'With land degradation, both local and global benefits depend directly on maintenance of the natural-resource base, so untangling the two to establish baselines is inherently difficult' (GEF, 2005a, p53). Moreover, it is not just the geographical application of benefits that are bundled; their biodiversity, watershed management and carbon sequestration are tangled together in such a way as to make them near-impossible to separate. This is discussed later in this chapter, when the broader PES approach is described.

But, in practice, there is a tendency with GEF projects to be pragmatic when defining incrementality. The distinction from the baseline does have to be made – but not with quite the same precision, and for these projects, a control-plot approach would not always be needed. With a CDM or similar project, it probably would. However, in both cases, additionality will have to be defined by an elusive standard – common sense.

Even so, baselines must be set to establish additionality. This is especially difficult if, as in the case of Kyoto and the CDM, baseline data has to be drawn for a year now past, for which historical data are either unavailable, or are misleading, or both. Poussart et al (2004a, S424) ask whether credits should be allocated simply on accumulation of SOC since project inception, or against a baseline of projected SOC levels without intervention. In the former case, scarcity of historical data would not matter, but the definition of additionality under the Kyoto flexible mechanisms demands a formal baseline. Setting these for soil carbon is difficult; as Izaurralde et al (2001, p556) point out, carbon stocks may not have been at a steady state at the time of initial measurement. The setting of baselines may therefore demand historical data over two or more fixed periods prior to project inception. Were they not to do so, however, it would be impossible to set baselines that reflected and rewarded prevention of SOC *loss*, as well as accumulation, making soil-erosion control a meaningless strategy. As Poussart et al (2004a, S424) say: 'In the cases where projects mitigate future loss of carbon from the soils, net changes over time could be … zero.' So the challenge must be met.

Moreover, as Noble and Scholes add (2001, p17), in LULUCF projects, baselines will have to be calculated in such a way as to accommodate year-on-year variations. Agricultural production is not constant. Environmental changes may also determine NPP – as Smith (2004, p267) points out. 'Fertilization … by increased carbon dioxide concentrations or from increased nitrogen deposition might increase carbon stocks but are not regarded as human-induced activities,' he says, and suggests that ecosystem models may be a way of separating the factors.

An alternative way might be stratified accounting with verification of agreed management practices and, crucially, with control plots, so that the baseline is actually validated during the projects. This might also simplify SOC monitoring and verification, discussed later.

Leakage

Mitigation projects are also expected to demonstrate non-leakage. Leakage occurs when, by mitigating GHG emissions in one place, a project causes them elsewhere. An example might be a fuel-efficiency initiative that reduces local consumption of coal so much that its price drops, and it is then used outside the project instead of natural gas. Another might be logging control that leads to deforestation in another location; as Ringius (1999, p28) puts it, shifting the problem around instead of solving it. 'However,' Ringius argues, 'soil carbon sequestration systems are unlikely to create leakage effects because they will frequently be more desirable than alternative land-use systems.'

This will often be true, but not always. First, a minimum- or no-till system, if successful, will eventually lead to better yields but in the first year or two they may drop as pests and diseases are dealt with, and local prices for cereals may rise, leading to less sustainable land use elsewhere. The same will also apply to any system of sustainable rotations; even if profitable at the farm-gate prices obtaining at project inception, those may change locally as a result of the project, perhaps leading to monocropping or irrigated wheat production on unsuitable land adjacent to the project area.

Reduced-tillage systems also demand retention of crop residues – and biochar projects could demand that they be used as feedstock. Either could cause leakage. In commercial farming and in areas with 1000–2000mm annual rainfall, such as southern Brazil, biomass production may be high and the opportunity costs of this may be insignificant. But in subsistence farming, especially in arid or semi-arid areas where overall biomass production is low, crop residues may have other uses.

In particular, they provide feed – to the extent that in an integrated crop/livestock system the grazing of livestock on cereal stubble may be economically more significant than the grain production itself. In some regions lentil straw can be worth more than the grain, and Saudi Arabia has actually imported it (FAO, 2002b, p6). In Syria, failure to appreciate this led to the release of at least one barley variety that, despite being high-yielding, was not adopted because the sheep found the straw unpalatable (ICARDA, 1995, p1).

Even where farmers do not graze livestock on their cereal stubble, they may permit others to do so, or sell the straw off-farm. This is in fact an important safety net for pastoralists on nearby steppeland, who are vulnerable to year-on-year variations in rainfall. In their study of Syria's Khanasseer Valley, which is marginal for arable farming and borders the steppe, Nielsen and Zöbisch (2001, p144) comment that land-use systems in dry regions are 'generally characterized by the coexistence and interaction of agriculture and pastoral activities ... [including] systems based on the exchange of different resources and products ... between two principally independent production systems'. In this case, if settled farmers retained crop residues instead of selling them, this could cause pastoralists to overgraze fragile steppe, causing substantial mineralization of SOC – that is, leakage.

The ways in which biomass is used on- and off-farm are, however, variable and complex. Bationo et al (2007, p14) report that in the Sahel, where biomass production is relatively low, much does get used for fodder – up to 50 per cent of millet stover, for example – but that other uses include construction and fuel for cooking. They report a study in the Sahel that found 90 per cent of crop residues being used for cooking (Bationo et al, 2007, p21). 'This practice results in considerable loss of carbon and nutrients such as nitrogen and sulphur,' they comment. Burning of crop residues in the field can also result in the apparently profitless emission of CO_2 – although there are usually economic reasons for doing it, as with burning of barley straw in Syria and sugar cane in Brazil. Thus, retaining crop residues in the field might cause leakage in some cases, but might actually have a double carbon benefit in others, depending on what the alternative use was (feedstock for biochar, perhaps) and how it was replaced. So the flow of biomass through the farming system would need to be modelled carefully when planning any project. These matters have also been raised recently by Giller et al (2009) in an important paper on conservation agriculture; the paper is considered as part of a discussion on tillage regimes in Chapter 8.

Leakage may also occur if a project demands intensification of production, and the environmental costs of extra inputs are not accounted for. Ringius (2002, p489) does raise this: 'N-fertilizer application, animal manure, crops residues, and nitrogen derived from N-fixing legumes could contribute to an increase in emissions of nitrous oxide (N_2O) ... from soils.' Ringius also points out that the emissions associated with the manufacture and transport of fertilizer could counterbalance carbon sequestration (Ringius, 2002, p489). If so, the use of fertilizer as part of sinks projects – especially for avoided deforestation, as Vlek et al (2004) suggest – might not make sense. Lal (2002, p11) breaks down the costs of inputs in terms of carbon; for example, 820kg/Mg C for nitrogen fertilizer. He also notes (as does Ringius) that the use of irrigation will produce carbon by increasing biomass but will have emissions costs of its own. They could also have mentioned N_2O emissions from fertilizer after application.

But three points should be noted with regard to this type of leakage. First, it is not hidden; these inputs are calculable and can be set against the project's carbon budget.

Second, agriculture in the developing world emits less carbon per hectare from inputs anyway, so there will be less to offset. Jung et al (2005, p26) report that these emissions are about 0.1t CO_2 per million calories (at point of consumption) against 1.7 to 2.2 tonnes in five developed countries.[1] Any percentage change in carbon emission from inputs is therefore likely to be small.

Third, recommended management practices for carbon sequestration are as likely to *reduce* emissions from inputs as to increase them. Thus eliminating tillage altogether would also eliminate fuel usage of 12.4 litres per hectare for a mouldboard plough and lesser amounts for other operations (Lal, 2002, pp355–356). Evers and Agostini (2001, p4) reported that in Brazil's Rio Grande do Sul province, mechanized farms benefited from savings of up to 66 per cent. This can be directly expressed in terms of emissions savings; thus Lal (2004a, p1624) states that where tillage with a conventional mouldboard plough results in losses of 15kg/ha C, chisel ploughing will lose 8kg/ha and subsoiling 11kg/ha. 'Therefore, conversion from conventional till to no-till farming reduces emission by 30 to 35kg/ha C per season,' he states. Presumably these figures would be subject to soil type, condition of machinery and other variables, and they are anyway not high compared with the balances from SOC, but they can be factored in.

This is not to say that calculation of emissions against benefits will always be easy. A harder case might be the use of inorganic fertilizer to increase biomass production in tropical pastures, where the cycling of nitrogen through the system is not always clear and where it may be altered drastically by changes in stocking rates. This subject will be raised again in a future chapter. Even so, the leakage from inputs will not always be a serious constraint to agricultural sinks projects.

But the other types of leakage – for example, from unmet demand for crop residues – might be, and must be addressed as part of independent verification, ensuring it is incorporated in the project design.

Permanence

The project design must also deal with the question of permanence. Mitigation projects that cut emissions – from a power plant, or a hospital, or by modernizing transport – are unambiguous; the carbon would have entered the atmosphere, but did not. Sinks projects, however, remove carbon from the atmosphere only for as long as it is not disturbed and mineralized, and this makes it hard to certify retention of sink carbon as an emissions reduction in the normal way. As was seen in the last chapter, there are pragmatic approaches to this, with the VCS certification scheme simply withholding a 'buffer' percentage of the tonnage against the sink's destruction (and also taking a less formal approach to baselines). But to register CERs rather than VERs, more rigorous techniques will be needed.

García-Oliva and Masera (2004, p352) argue that SOC has an advantage 'over sequestration in above-ground biomass, because the risks of partial or total removal of the carbon benefits are generally lower'. They are presumably thinking of forests. But agricultural sinks could involve above-ground biomass, and besides, SOC could certainly be lost. Izac (1997, pp265–266) sees farmers regarding SOC as natural capital that they will use in the role most profitable for them. That role can change. Blaikie (1985, pp45, 85) has described how heavy Soviet grain purchases led to American farmers ploughing some 27 million hectares and wiping out the gains from years of soil conservation programmes. As the GEF official referred to earlier in this chapter commented, 'If farmers need to use up the carbon, they'll do so.'

Nonetheless, a tonne of soil carbon mitigates climate change for the time it does remain in the sink. As Fearnside (2002, p19) states: 'Credit given for options that result in temporary carbon storage should not be equal to that from permanent displacement … but neither should it be zero.'

In effect, then, this is an accounting problem. McCarl and Murray (2001) take the hypothetical case of a no-till project under which soil carbon saturation is reached after 20 years. If the farmer ploughs at the end of this period, the carbon is worth less than an emissions reduction from, say, a power station, because the latter never entered the atmosphere but the SOC eventually did. If one calculates the radiative forcing potential of the CO_2, it is possible to work out how much less the temporary storage – the agricultural sink – was actually worth.

In this case, McCarl and Murray say that the agricultural sink was worth 38 per cent of the power-station reduction. If one accepts their figure (and the calculations would be complex), then, assuming a carbon price of $5 per tonne, one might pay $100 for avoiding emissions of 20 tonnes of CO_2 from a power station over the same period, but should only pay $38 for the sequestration of that carbon in agricultural land. However, if the agricultural sink were maintained under subsidy so that it was not released for 100 years, it could be worth 55 per cent rather than 38 per cent of the emissions reduction (McCarl and Murray, 2001, p7).

And there is a third scenario under which it is maintained by the farmer for his/her own purposes. McCarl and Murray admit that in such a case, a sequestration offset might be worth *more* in mitigation terms than an emission reduction. 'In this case,' they say, 'the payer of the offset receives a free ride [because more carbon has been kept out of the atmosphere than the investor paid for]. However, such effects could also occur through an emissions reduction program (e.g., a technical change subsidy) which would continue after program end thereby also generating mitigation benefits beyond the payment period' (op. cit., p11). In these cases, the value off the offsets are theoretically greater to the investor, although that would probably not be internalized in the transaction.

A separate but related issue, highlighted by García-Oliva and Masera (2004, pp352–353, 357), is the fact that SOC sequestration may take anything between 20 and 50 years to reach saturation, whereas the accounting periods for CDM projects are short; those for other mechanisms may be longer, but are unlikely to accommodate contracts for 50 years. And they will definitely not be contracts in perpetuity.

The answer to all this is to not 'buy' carbon, but to 'rent' it; that is to say, instead of paying for a tonne of carbon on the assumption that it will never be emitted, and then retiring the credit against emissions, the buyer would pay a smaller sum periodically in order to keep the tonne out of the atmosphere *for that period*. The buyer need not worry about what happens to the carbon at the end of that period because s/he is not paying for its retention in the sink beyond that.

There is a problem here in that the farmer would have an incentive to plough and release the carbon at the end of the contract, and then negotiate a new one. In practice, however, this might not happen in agriculture, as the soil carbon content will take many years to reach equilibrium, so will continue to increase if left undisturbed. Because this has agronomic advantages, there is a good chance the farmer would let this process continue even if not receiving carbon credits for doing so. This would continue until the SOC content reached equilibrium, a point that could be determined in advance using a model such as SOCRATES[2] or Century (these are discussed further later in this chapter). At this point, the farmer would, in theory, have an incentive to release the soil carbon, as s/he would have to do so in order to negotiate a new contract. But doing so would destroy the benefits of SOC to soil structure and productivity, which should by then outweigh the market value of the carbon, unless the market has risen very high.[3]

This does assume that Fearnside is right, and there is mitigation value in *temporary* sequestration of carbon. Dutschke (2002, p383) quotes Chomitz's argument that there is, on several grounds. First, it gives more time for biotic systems to adapt to climate change. Second, every unit of atmospheric carbon will eventually decay, so is itself temporary; it is therefore useful to offset with temporary, as well as permanent, emissions reductions. Third, slowing climate change lowers the net present value of the damage it does, both because of discounting against interest rates – as in any economic process – but also because future generations will be better able to adapt. Fourth, Dutschke argues that future emission of a tonne of carbon will have a lower marginal value as it will be released into a larger atmospheric carbon pool. Fifth, the cost of abatement technologies will be lower in the future.

These points can be challenged. The first – that slowing or delaying emissions gives biotic systems time to adapt – is probably true (although it is people's relation to biotic systems that needs to adapt, as much as the systems themselves). Regarding the second point, that GHGs in the atmosphere themselves eventually decay, Dutschke does acknowledge (Dutschke, 2002, p382) that the six GHGs have highly variable decay times and that although they are as little as 12 years (for methane), they are 'bundled' with those that decay far more slowly, implying 'near eternity' of sinks. But theoretically it is quite correct, as ongoing sequestration can be matched against the ongoing decay. In any case, the GHGs could be unbundled for accounting purposes. The third and fifth points are essentially similar, as adaptation and better mitigation are part of the same process, but they are probably true in principle.

But the fourth point – that a tonne of emissions will have a lower marginal value if emitted later – may not be true at all. If a tonne of carbon is emitted into a larger carbon pool, it may hit an ecosystem that is far nearer a tipping point than it would have been at a lower temperature; the potential for sinks to flip into sources as the temperature rises was discussed in Chapter 2. And in general it seems that these points are theoretical. They are useful debating points, but the real reasons for renting, rather than buying, carbon are pragmatic. One is creating what will in at least some cases be permanent sinks, but cannot guarantee that they will be, so it makes more sense to contract for short periods rather than permanent removal of carbon.

Dutschke proposes two accounting mechanisms. The first, tonne-year accounting, incorporates an 'equivalence time' after which the damage caused by the emission of a tonne is counterbalanced by sequestration. This is put at 55 years, though it is stated that the choice of a timeframe is political as much as it is scientific. As Dutschke points out, this type of accounting can also be used to calculate remaining liabilities in the event of a project that ends ahead of time.

There is some complexity (for example, the decay time of a tonne's climate impact is non-linear, and Dutschke therefore suggests that the 55 years be divided into three linear fractions to compensate). Nonetheless, it has important attractions. The problem of permanence is solved because, after 55 years, a given tonne of carbon's effect on climate actually *has* been removed, even if the mathematics is open to argument. Moreover the farmer is paid according to the global benefits s/he has or has not provided, no more, no less; it thus has legitimacy.

The alternative to tonne-year accounting is a simple rental, whereby the investors offset their emissions against a carbon pool for a given period; at the end of that period, assuming their emissions continue, they must either renew the contract or offset them elsewhere. This is called a temporary CER, or T-CER (Dutschke, 2002, p387). There is an advantage here over tonne-year accounting, in that, under the latter system, there would be no incentive to maintain the sink after the 55 years had passed. The tonne in question has been offset. A T-CER, however, can be entered for a new rental period. However, there are some problems with this. One is that it is impossible to fix the real value of a T-CER without knowing what its fate will be after the end of the contract period. Moreover T-CERs give a tonne of carbon a potentially unlimited validity; there is no timescale. So a given tonne of carbon could end up being used, over several rental periods, to offset an awful lot more than a tonne of emissions (the CDM has provisions to prevent this in LULUCF projects; other mechanisms might not).

Dutschke's answer to this (2002, p389) is a leasing scheme whereby an investor may buy T-CERs that have a declining offset value over a 55-year timeframe. At the end of that period, to be sure, the land user will have no further incentive to maintain the sink, at least in theory.

This is important. If they do not, non-permanent accounting for sinks may cause emissions, not save them. Meinshausen and Hare (2002, pp1–7) argue persuasively that if a tonne sequestered is used to offset a tonne of emissions from (say) industry, but is itself subsequently emitted in later years, then we have lost more than we have gained. But they add: 'Of course *permanent* carbon storage that is additional over all time would have the same effect as reduced fossil fuel emissions, but … this is an unlikely situation.'

In agriculture, it might not be. Its advantage over forestry should be that soils with a high SOC content function better than those without, as the high organic matter content results in less erodibility and greater productivity. It is also hard to be so sure over decay rates under different emissions scenarios. But even if Meinshausen and Hare's view on temporary credits is not accepted, it should be borne in mind. If permanence of sinks *is* solely an accounting problem, then Dutschke has solved it. But sinks should only be accepted if there is a reasonable chance they will be permanent, even if the contracts are not.

Monitoring and Verification

The need to monitor and verify on-farm carbon sequestration is, along with baseline-setting, one of the main constraints to agricultural sinks. There is a view that it is a technical issue and can therefore be solved. The author has encountered this view more than once in discussions with officials; a classic case of the epistemological position 'get the science right, then act'. However, the application of science to policy is not always so simple, and the monitoring and verification of agricultural sinks is a complex problem. This section will argue that it can be overcome by combining different methodologies. However, the difficulties should not be underestimated.

A major part of the constraint is measurement of soil carbon. This is not because it cannot be done; an accurate method, the Walkley-Black procedure, has been available since the 1930s, although it is sometimes now being superseded on environmental grounds because it measures oxidizable SOC by digesting it with chromic acid, raising questions of safety and disposal (Kimble et al, 2001, p7). However, it remains widely in use. A more modern practice is weight loss on ignition: the soil sample is heated in a furnace and the CO_2 given off is calculated according to the sample's loss of weight. This is established technology. 'Contrary to general perception,' says Lal (2002, p359), 'SOC can be measured accurately, precisely and in a cost-effective manner.'

But there are some practicalities when matching the available technology to soil carbon projects. Izaurralde and Rice (2006, p460) point out that the short-term changes that might need to be detected might be very small (perhaps 0.1–0.5kg/m^2 C) in relation to the stocks present, which might be (say) 2–8kg/m^2 C. This means that techniques that have proved quite satisfactory when used less frequently and for other purposes may prove uneconomic for monitoring and verification of a sink.

For example, Lal (2002, pp358–359) gives ten steps for assessment at field scale. They include preparation of a detailed soil map (≤1:50,000 scale); soil sampling at several depths; sub-samples for each depth and composite, with three profiles for each land use/soil management treatment; aggregate analyses; and determination of SOC content in sand, silt and clay fractions. This might not be difficult in developed countries and, as Lal points out, the results can be scaled up across similar soils and ecoregions. In the Great Plains of North America, these could be large. However, in developing countries with inadequate extension and laboratory facilities, these requirements could be unrealistic, especially where the findings must be scaled up across heterogeneous landscapes and cropping systems.

Yet it is not easy to eliminate any of these steps and still obtain an accurate result. For example, Lal includes measurement of bulk density. Knowing how much carbon is in a given sample is not meaningful if one does not really know how much soil it contains, as carbon payments will be based on the number of tonnes, not the percentage of SOC (Kimble et al, 2001, p9). Bulk density can be subject to rapid change through compaction by traffic and by agricultural processes; these would include the management practices implemented to sequester the carbon. A related requirement in Lal's list is aggregate analysis. Again, this is necessary to confirm the SOC content of each aggregate fraction, as the different carbon pools will have very different turnover times (Six and Jastrow, 2002, pp939–940; this was discussed in Chapter 2, above).

There must also be sufficient samples to allow for spatial variability. Conant and Paustian (2002) say that this can be 'substantial … with coefficients of variation [CV] as high as 20 per cent even in a visually uniform cultivated field'. Moreover, this variation is not predictable. Campbell (1979, p546) argues: 'There is hardly a soil, or a class of soils, for which we can answer the question: "How much variation can be expected for soil property *A* within a distance of *X* meters?"' As the CV rises, the number of samples needed to estimate the mean within a given confidence level increases exponentially. According to Wilding et al (2001, p75): 'It should be noted that to achieve an accuracy of ± 10 percent of the population mean for any given CV requires about four times as many observations as for ± 20 percent.'

It does not help that year-on-year changes in SOC may be limited, to the extent that they may be small relative to the standard error involved in measuring them. Jones et al (2007, p53) state that if measurement error of SOC is 0.04 per cent on a mass basis, computed measurement standard errors of about 1000kg/ha/yr C may be expected; they point out that Lal himself reported SOC build-up in a no-till experiment in Nigeria at 400kg/ha/yr, less than half that amount.

From Sampling to Modelling

These difficulties may need to be overcome by a fresh perception of the problem. Kimble (2006) argues that there is a tendency for soil scientists to:

> *focus on finding variation … [I]n soils we focus on a level of precision and accuracy that may not have any relevance to the real world … We as soil scientists need to … explain the variability … and not use it as a crutch for not wanting to answer policy questions.* (Kimble, 2006, p488)

Wilding et al (2001, p84) take a similar view, and argue that understanding the *sources* of variability can help provide reasonably accurate estimates, although uncertainty cannot be eliminated.

Besides, projects over long timescales may not need frequent verification; Cheng and Kimble (2001, p126) point out that year-on-year variations may be very small or even negative, and it makes more sense to model them as averages over a 5- or 10-year period. In any case, they say, spatial variability is no argument for excluding agricultural SOC from the Kyoto Protocol when forests *are* permitted: '[T]he same variability … occurs in any estimation of the carbon contents of the forest biomass. … The variability in the biomass of trees is probably not any less than that of soils' (op. cit., p123). Cheng and Kimble are satisfied that measurements can be scaled up: 'By using appropriately selected point measurements and modelling, carbon values for a farm, a region, or even a country can be estimated.' Conant and Paustian (2002, S135), discussing grasslands, also believe that modelling can decrease sampling intensity (although this conclusion is based partly on the existence of data that might be unavailable in lower-income countries). So, given sufficient accuracy in modelling, the recommendations in Lal (2002) need only be followed at the beginning of a project, for calibration purposes.

Can SOC be modelled sufficiently accurately for certification of emissions reductions? If so, which is the best model? Izaurralde et al (2001, pp553–575) tested six such models. Two models (including the well-known RothC) were excluded from the test at an early stage, based on 'ease of use, input requirements and run time' (op. cit., p559). The remaining models were tested against long-term data from sites in the Canadian provinces of Alberta and Saskatchewan, covering different combinations of cropping, tillage and fallow regime. The authors selected SOCRATES as the most accurate overall, with Century and two others not far behind. Century has higher input requirements than SOCRATES but plant growth simulation includes trees – perhaps not needed in the cereal-growing regions of North America, but essential in the agroforestry systems that are more likely to be used in carbon projects in the developing world. It also includes different tillage regimes, multiple crops per year (likely in the tropics) and unlimited rotations, and CO_2 fertilization. Moreover, the models tested did not include the US Department of Agriculture's recent COMET-VR 2.0, which is effectively an improved way of running the Century model, and accounts for a range of factors including N_2O emissions and fuel use.

Century has been used with success by Poussart et al (2004b) to simulate soil carbon changes in the semi-arid zone of Sudan, despite variable historical data. The authors do not claim pinpoint accuracy; rather, they say that such models can 'be a valuable tool to simulate SOC changes resulting from … improved land-management practices in the Sahel …

Even if estimating historical or current SOC levels is somewhat problematic, modelling changes in SOC from a given baseline may be accurate.'

This last sentence is perhaps the key to the way in which agricultural sinks should be measured and verified. The demands of accurate farm-level measurement, as outlined by Lal (2002, pp358–359), will not be cost-effective for small-scale, heterogeneous plots, especially where the baseline level of SOC is low. It will be easier to model probable SOC changes in Century against baskets of management practices, and then verify that these have been implemented.

This need not be done solely on the ground. Remote-sensing images from satellites or aircraft can be used to periodically monitor random plots for management practices and biomass. In any case, a narrow focus on measuring SOC may not address the issue. Agriculture sequesters large amounts of biomass, much of which is above ground; it will be mineralized back into CO_2 in time but is still part of the carbon budget, so if a farm has 100 tonnes of above-ground biomass for six months, this is equivalent to holding a smaller amount for more of the year. Besides, the continuous cropping cycles on some tropical farms mean that there will be an above-ground sink most of the time. So although SOC, with its relative permanence, is the important component, one should still look at total system carbon.

Woomer et al (2001) present a possible approach arising from a study in smallholdings in the Kenyan Highlands. 'Methods for characterizing carbon stocks and flows in smallhold farming systems are not well established and the size, distribution and opportunities for increasing farm carbon stocks are insufficiently documented,' they say (op. cit., p147). They provide a step-by-step guide for measuring woody biomass, herbaceous vegetation, surface litter, roots, total soil carbon, and those labile fractions most susceptible to management changes. In each sampled farm, a feature (perhaps a bare field) should be selected and fixed using a geographic positioning system (GPS). The authors accept that it can be difficult to interpret Landsat images, as 'each pixel may contain complex mixtures of structures, trees, crops and bare land'. But remote sensing is a work in progress, and the resolution available is likely to improve.

It is possible to envisage an approach that combines Century with Landsat or aerial-photography images to give annual average total system carbon, including SOC; control plots can be provided with which to validate the baseline continuously during implementation. Given this lower emphasis on meticulous soil sampling, monitoring and verification of agricultural sinks may be easier than it appears. Pinpoint accuracy may not yet be feasible at plot level, but where payments are predicated on the adoption of certain management practices, these can be verified on the ground. This is more or less the approach used by the PES projects discussed at the end of this chapter.

However, a major question remains: monitoring loss of SOC through soil erosion.

Measuring Erosion: Is it Possible?

It is hard to see how this question can be avoided. As discussed in Chapter 2, the global total of SOC emitted as CO_2 following translocation by erosion could be 0.8–1.2Pg/yr C, the equivalent of about a third of the annual increase in CO_2 (Lal, 2003, p437). As was explained, this view has been challenged, but even if erosion is eventually proved to be a net sink on a global scale, this may be far from true of individual watersheds in project areas. The implication is that these losses do have to be modelled or, failing that, measured. Techniques do exist for measurement (for example, use of the isotope ^{137}Cs) but this would not be practical on a large scale.[4]

Post et al (2001, p78) state that erosion 'can confound efforts to monitor changes in soil organic carbon storage'. Measurement is, they say, 'involved and costly' and likely to remain in the research domain. They suggest the use instead of simulation models such as the Revised Universal Soil Loss Equation for carbon sequestration projects. 'Most likely, erosion estimates with simulation models, together with information on carbon content in sediments and sediment enrichment ratios, will be the most practical way to estimate carbon changes caused by erosion and deposition' (Post et al, 2001, p78). But as Stroosnijder (2005, p163) points out, models must be calibrated, so measurement is needed. The only workable approach would probably be to use a proven technique (say Gerlach troughs) on selected plots in areas where erosion appears widespread. A rough estimate can then be made of the amount of lost topsoil, and the carbon budget amended accordingly.

There are some problems with this. First of all, although an established model, the Universal Soil Loss Equation (USLE), does exist, its reliability has been questioned when applied to heterogeneous tropical landscapes with steep slopes. Although there have been attempts to adapt it (there is a Revised version, or RUSLE, that Post et al refer to), it is of limited use with narrow bench terraces and steep hillsides, according to Critchley (2000, p150), who says of the USLE that 'its universality is a better description of where it has been applied than its relevance: it is frequently used out of context'.

Even were there an alternative model, there would be further challenges in obtaining data to calibrate it. Sediment trapped for the purpose may be of unclear origin, especially where wind erosion is a problem; in the case of soil transported by water, the source may not be known if the test plot is unbounded, but if it is not, there will be no inflow into the plot (Stroosnijder, 2005, pp164, 167). There are also specific environments which may be problematic: for example Critchley (2000, p84) highlights the lack of a reliable method for quantifying runoff and erosion from certain types of terracing. And there are more general problems; Stroosnijder (2005, p171) points out that erosion measurements are expensive and therefore sometimes restricted in time and space, so they don't allow for inter-annual or intra-seasonal variations, while upland erosion may be masked by bank erosion in a stream.

Stroosnijder (2005, p171) also highlights problems with consistent and adequate construction, handling and calibration of equipment. He is not the first to do this. In the first issue of *Land Degradation & Rehabilitation* (now *Land Degradation & Development*), Carpenter (1989) warned that uncertainties are especially prevalent in reports from developing countries:

> *The difficulties of field research in developing countries are enormous. ... Logistical problems include bad roads, high corrosion rates on equipment, absence of spare parts, hostile local inhabitants and language/literacy difficulties. Sheet metal and plastic devices installed at the site disappear, 'requisitioned' for other uses. A typhoon washes out the apparatus and a laboratory has no electricity for two weeks while samples deteriorate. Gauging stations are unmanned during storms when the most valuable data are to be obtained. ... When data from the host nation are lacking, then information from other climates, countries and cultures is applied and further uncertainties are introduced.* (Carpenter, 1989, p1)

One might not always be thanked for pointing all this out. But there are real difficulties in measuring soil erosion accurately, and in developing countries they may indeed be exacerbated by logistical constraints, and any carbon-trading paradigm must confront these if it is to be credible. The scale of the problem is reflected in the titles of two papers cited here; Stroosnijder's (2005) is called 'Measurement of erosion: Is it possible?' while Carpenter's (1989) paper is headed 'Do we know what we are talking about?'

This would seem to confound attempts to tie soil loss to specific farmers for accounting purposes. This is an area where high technology has not so far helped much; besides measuring ^{137}Cs, there is increasing potential to use remote sensing, but this is also limited in what it can do. While it can certainly supply information about land-use that can be used to model erosion, there is no proven way of using remote sensing to *directly* quantify soil erosion, other than the extent and development of gullies (Stroosnijder, 2005, p165). Perhaps the most practical approach is to take plot-level measurements using very simple techniques such as erosion pins on sloping plots. As Stroosnijder says (op. cit., p167), farmers themselves use these, and they can measure both wind and water erosion through changes in the distance from the top of the pin to the surface. Erosion pins should not be used for measuring an individual farm's carbon budget; they can be tampered with, and besides any rise or fall in surface level may be due to activity in a neighbouring plot. However, the technique can be used to measure the soil that has actually been lost from that plot for calibration purposes.

To know how much carbon has moved with it, it will be necessary to know how much SOC was there and, importantly, where it was in the soil profile. However, advances in spectroscopy mean that this may soon become economic; these developments are discussed below. It may then be possible to combine this data with that from sediment traps in watercourses and establish roughly how much carbon is leaving the watershed as SOC, and how much is being emitted as CO_2 due to erosion. This will allow reasonably accurate erosion/carbon budgets for a given watershed and, in time, what is learned will permit better modelling of the relationship between erosive processes and CO_2 emissions from farmland.

At plot level, however, it may have to be accepted that there is no completely fair and accurate way of doing this for now. It may be better to model average rates of erosion, along with other factors affecting sink size, across relatively homogeneous areas according to land use – the latter being determined by remote sensing. Doraswaimy et al (2007) tested such a method in Mali. However, even in this relatively homogeneous area, overestimates of carbon sequestration were possible because previous studies had used lower-resolution Landsat imagery that could not resolve smaller field sizes. As a result, non-cropped land had sometimes been included in the total (op. cit., p70). Remote sensing and modelling can clearly work over large areas, but this is a warning that there is a trade-off between sampling resolution and accuracy with remote sensing, just as there is with sampling on the ground.

Alternative Sampling Methods

Approaches such as modelling and control plots will help reduce the dependence on traditional techniques such as Walkley-Black or weight loss on ignition, but there will still be a need for ground-truthing during project implementation; and although these techniques will probably always be needed to calibrate models, they will be too costly for ongoing monitoring and verification, especially in the developing world. Reeves et al (2006, p424) report estimates that 'tens to hundreds of millions of samples may need to be analysed yearly for Brazil alone'.

The answers may come from outside agriculture. Izaurralde et al (2001, pp554–556), discussing aspects of carbon measurement, see a need for 'field instrumentation necessary to make fast and reliable in situ determinations of soil carbon across varying landscapes', but see opportunities: 'Instruments that map soil salinity by induction principles (EM38) or determine soil water content by time domain reflectometry (TDR) provide excellent examples of the advantages from developing in situ technologies' (op. cit., p556). Of the technologies they cite, one was designed for purposes outside agriculture, while the other has since been used for them. (TDR was developed to locate shorts in inaccessible cables by measuring a reflected pulse. The EM38 is mainly used to measure soil salinity, but has also been used in archaeology.)

This sort of cross-fertilization may uncover the technologies needed. near infrared reflectance spectroscopy (NIRS) has long been been used to establish nutritional components of feed and forage (Ludwig and Khanna, 2001, p361); for crop-breeding purposes, Consultative Group on International Agricultural Research (CGIAR) researchers have used it in this way to establish probable feed quality of barley breeding lines

(Goodchild and Jaby El-Haramein, 1996, p5). Ludwig and Khanna (2001, p361) explain that it is not always possible to assign a part of NIR spectra to chemical characteristics. 'However, by relating the results of chemical analysis to the NIR spectra of a large set of samples, the composition of soil and litter constituents can potentially be determined indirectly,' they continue.

This raises the attractive possibility that NIRS might be used not only to determine total SOC, but to identify different carbon pools that might be expected to have different mean residence times (MRTs), a process that might otherwise demand aggregate analysis and/or a detailed cropping history (Ludwig and Khanna report that experience with this has so far been mixed). NIRS is unlikely to replace Walkley-Black or weight loss on ignition, but as a tool for periodic ground-truthing and calibration of models against Landsat data, it is very promising. And Reeves et al (2006, p439) state that near- or mid-infrared spectra can also be used to determine not only total carbon but also nitrogen, pH and other characteristics. Build-up of carbon is very closely related to these characteristics; for example, in tropical pastures it can be closely related to nitrogen cycling, which can itself be greatly changed by stocking rates. Reeves et al also suggest that the ability to determine multiple characteristics may assist discrimination between labile and non-labile carbon. This, they say, would be 'very informative in indicating how much of the carbon found in a soil sample would be likely to be sequestered for any length of time' (Reeves et al, 2006, p433).

Other tools with potential (although possibly at a cost) include the use of lasers on samples and the spectroscopic analysis of the light that they then admit. Known as laser-induced breakdown spectroscopy (LIBS), this is said to be potentially field-portable, allowing soil-carbon measurement in real time, and could also give differential measurements of carbon content across the soil profile (Ebinger et al, 2006, pp408, 416–417). The latter quality would be very useful not just in carbon projects but in research into the mechanisms of soil carbon accumulation. Ebinger et al suggest that potential for simultaneous soil nitrogen and carbon analysis to be done in the field 'is within current field-portable LIBS technology'. If economic, this would be an interesting option. Also intriguing is inelastic neutron scattering (INS), described by Wiepolski (2006, pp433–455). This demands an initial investment of about $100,000 plus some periodic costs, but is non-invasive and could be used in a continuous mode over large areas. However, given that NIRS is already used for related tasks such as feed quality, it may be that it is the most practical of these technologies.

Besides actual soil analysis, flux measurement also has a role. Post et al (2001, p81) explain that eddies – vertical air movements over a vegetated surface – can be measured and correlated with CO_2 to calculate net uptake and release of carbon from soil and vegetation. The accuracy of this method has improved greatly over the last 20 years (ibid.). There are several basic methodologies. Costa et al (2006, pp696–697) describe how a simple chamber can be constructed and air samples taken; these can then be analysed for CO_2, N_2O and CH_4 using gas chromatography. Alternatively, infrared spectroscopy can be used.

Flux measurement does measure fluxes from fixed points and would presumably involve the same trade-offs between density of sampling and accuracy as traditional soil sampling. Smith (2004, p265) suggests sampling using tall towers or balloons for landscape or regional-scale budgeting or even flask measurements and flux measurements from aircraft for continental-scale budgets. For areas of less than a square kilometre, eddy covariance can be used (ibid.). Typically this will be done with a Bowen ratio system, which looks very much like a small weather station. The technology is proven. In the late 1990s, Uzbek, Turkmen and Kazakh researchers began working with the International Center for Agricultural Research in the Dry Areas (ICARDA), several US universities and the US Department of Agriculture to monitor flux from rangelands near Samarkand in Uzbekistan (Nasyrov, 2000, p31). The project aimed, among other things, to assess the role of the vast Central Asian rangelands as a carbon sink and to investigate the effect of different land uses and management practices on CO_2 budgets, and to investigate to what extent the data can be extrapolated over a wider area. The field data, collected at 20-minute intervals, was transmitted to one of several collaborating American institutions and analysed in five-day segments. The data was encouraging, with 698g/m^2 sequestered during the growing season (Nasyrov, 2000, p32).

Flux measurement using eddy covariance has limitations. Post et al warn: 'The eddy covariance method requires some degree of continuity in air movement that may not be met under conditions of atmospheric inversions or very low wind speeds. Appropriate estimation of CO_2 fluxes during these periods is essential and requires considerable experience with the particular site' (Post et al, 2001, p82). However, even if the Bowen ratio system could not be used for verification, it might certainly be suitable for calibration of models at project inception in a given ecosystem.

Monitoring of SOC presents real difficulties, but it is possible. True, laboratory measurements of soil samples cannot easily monitor sequestration projects as such, chiefly because spatial variability of SOC requires too high a number of samples for reasonable confidence. What it can do, is calibrate models that can be used to project SOC against a baseline given certain management practices, with the latter verified using Landsat imagery combined with ground-truthing. Meanwhile the baseline control plot and project SOC levels can be periodically checked, albeit at a lower level of confidence than with lab samples, possibly by using portable spectrometers.

There are costs involved in all this, and skills will be needed, but generation of CERs for compliance is likely to demand these sorts of standards. What if they are not possible?

PES: An Alternative to the Market?

One possibility is to accept that the difficulties of baselines, monitoring, verification and permanence outlined in this chapter so far are just too difficult, and that formal cash-based trading for fixed units of carbon is not possible in agriculture. The author does not accept that; the problems are soluble, and may have to be solved anyway if we are to account accurately for terrestrial carbon stocks. In the meantime, however, a simpler alternative may be payment for ecosystem services (PES), which is coming into use in rural communities – there is now growing experience, especially in Latin America.

PES is a growing trend in international development. It seeks to compensate land users for managing the land in such a way as not to impose costs on others. Otherwise, they may have no incentive to avoid activities that affect others but not themselves; that is, activities that have negative externalities. Its disadvantage for our purposes is that it does not account so accurately for carbon stocks, so may not easily leverage funds from the private sector or get agricultural carbon onto the compliance market. It will therefore usually have to be funded by governments or by official development assistance (ODA), or by internationally agreed transfer mechanisms such as the GEF.

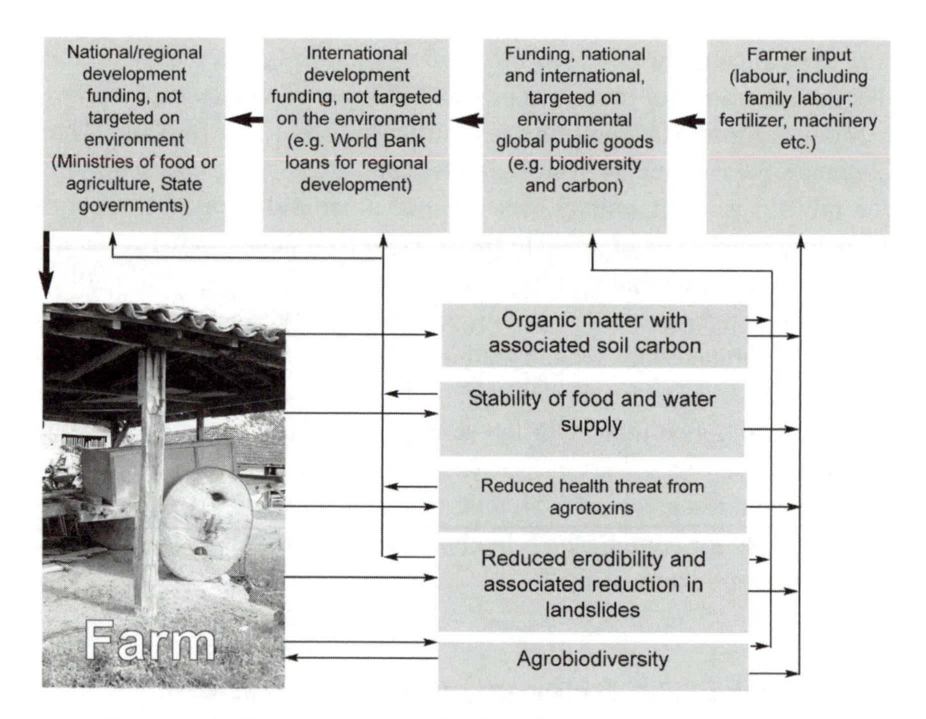

Figure 5.1 *Funding and benefit flows from environmental services projects in agriculture*

Its advantages are, first, that simpler accounting and payment systems are possible; this will be discussed below – and second, that it allows environmental goods other than carbon to be accounted for, at least loosely. Land degradation associated with unsustainable farming has serious negative externalities. GHG emissions from agriculture is only one of those; the classic examples would be those that reduce water availability and quality in nearby cities, perhaps through deforestation, soil erosion and runoff of potentially toxic chemicals. Thus soil translocated by erosive processes may block irrigation canals or dams. Flooding as a result of loss of water retention in hillsides is also an externality and is sometimes quantifiable. For example, in 1999–2000 in Kyrgyzstan nearly $2.37 million was spent on anti-landslide construction and bank-strengthening (IFAS, 2003). Siltation of watercourses is also a significant externality, and is frequently quantified in terms of crop losses or maintenance costs – or in accelerated depreciation of a capital asset; thus FAO (1994) estimates the expected life of eight Indian reservoirs as being from 23 to 79 per cent of their original design life, with four below 40 per cent.

Land-use measures to augment the carbon sink may have a bearing on other environmental goals. Cowie et al (2007a, p340) compare strategies for carbon sinks with their effect on biodiversity and desertification. This comparison includes reduced tillage, which they see as mainly positive for both goals (although they note possible increases in herbicide use and the impact on biodiversity), crop residue use (part of reduced tillage anyway, according to the IPCC; this is positive for both other goals), organic amendments (generally positive, but depends on contaminants), and irrigation (could increase biomass and prevent conversion of marginal land, but has potential negatives, in particular salinization). In general, they see more complementarity than conflict with the techniques used for agricultural sinks. In fact, Cowie et al (2007a, p349) argue that 'all major environmental goals should be pursued jointly, rather than independently as is the dominating current practice', and point out that land-use climate mitigation decisions taken in an integrated manner may not be the same as those taken piecemeal.

This is clearly true, but operationalizing any formal relationship between the major environmental treaties in agriculture would be very difficult. Biodiversity is one of the best examples. Biodiversity is a broad term covering all life forms, which of course include fauna;[5] steppe areas, for example, can contain wild animals that may be under pressure from expansion of grazed area, while those in forests may be threatened by extensification. Moreover, as discussed earlier, land degradation can force extensification of agriculture at the expense of marginal areas or forest that may contain plants with medicinal value or even wild relatives of crop species.

The value of these genetic resources is hard to estimate. Wild animals may have significant value to local communities; globally, however, they will have a mainly hedonic value. Plants have a more direct economic value and it is often global. Where they have been used for pharmaceutical purposes, there is a clear market, although this can lead to secrecy in the use of these resources (IPGRI, 1996). Genetic resources used for food production also, in theory, have a global value. This is because most regions grow crops using genetic resources that originated on another continent. According to IPGRI (1996), North America's food production is 100 per cent dependent on them, the Mediterranean region 99 per cent, Latin America 56 per cent and West Central Asia – the lowest figure – 31 per cent. Farmers who husband on-farm genetic resources are therefore generating a significant global benefit. But there are serious constraints calculating these values in cash terms.

First, as IPGRI (1996) points out, it is hard to relate benefits to whoever's farm supplied the genetic resource. It quotes the example of the rice variety IR36, which has 15 landraces and one wild species in its heritage and took 20 years to breed. The development of molecular techniques may permit the use of even greater diversity in crop breeding (Elings et al, 2000). To complicate matters further, such allocations would have to reflect the added value from conserving the genetic resources on-farm (*in situ*), rather than in a gene bank (*ex situ*), ensuring the continuing adaptation of the plant *in situ* (ibid.) – an especially valuable ecosystem service given projected climatic variability. But to whom, exactly, do these benefits accrue, and who provided them?

Where these multiple benefits cannot easily be valued, and the beneficiaries are hard to identify, PES can provide an informal, pragmatic alternative in which carbon conservation and sequestration can be bundled with other goals. The oft-quoted classic example is New York City's initiative with farmers in the Catskills watershed, which so improved the urban water supply that expensive water-treatment options were not needed (Salzman et al, 2001; Scherr et al, 2004, p33). Such schemes have an added urgency when farmers may be not simply unwilling but unable to address these questions without assistance, and in developing countries this may well be the case.

PES as a principle is not unquestioned. Dudley (2005, p3), for example, asks whether the landowners also own the ecosystem: 'Is the implication that, without payments, they can do what they want?' he asks. In fact, they probably cannot, at least in theory, as there is usually national legislation regarding land use. However, PES also implies monetary valuation of ecosystem services, a principle that is not universally accepted. These questions were discussed in more detail in Chapter 3. But the cash value of land in any country will depend to some extent on the services it is permitted to provide; PES is not conceptually different from the regulation of the industry that would be found within the EU, for example. And as Rosa et al (2004) point out, where agriculture has become marginalized, such schemes may allow rural people to reassert their importance in the development process.

That said, PES is effectively an incentive scheme, and there are some problems with incentives in agricultural development. This subject will be raised again with reference to the field survey described in Chapters 7 and 8. For the moment, it is worth looking at the place of on-farm carbon in PES so far.

Experience with PES

There are now a number of PES activities under way that have a bearing on the carbon sink. These are now being tracked as part of a potentially very important initiative, FAO's Mitigation of Climate Change in Agriculture (MICCA) project, which started in early 2010.

This project has tried to establish how many relevant projects are going on, where, and what they are. From this, Seeberg-Elverfeldt and Tapio-Biström (2010, p4) report that a number of projects were identified. They do not claim that their list is comprehensive (in fact, it excludes the three-country GEF-assisted project described below). However, they do briefly describe 21 projects that include a carbon sequestration objective. Some are really reforestation rather than agriculture projects, but there are several that are relevant to this book. They include a sustainable land management project for smallholders in Western Kenya, a major sustainable-grazing collaboration between FAO and China, and a project encouraging Egyptian farmers to use crop residues/compost for soil improvement.

However, most of the initiatives that Seeberg-Elverfeldt and Tapio-Biström report are fairly new, or are in the planning stage, or are just beginning now. They are also not, with one exception, seeking to formally register CERs (although some are planning to sell VERs on the voluntary market, three of them certified under the VCS). This will change, and the existence of MICCA will probably be helpful in itself. So far, however, there is very little experience with any formal carbon offset project in agriculture.

But there has been some with PES, notably in Latin America. There were also one or two early forestry projects under activities implemented jointly (AIJ) that involved diverse systems and had some bearing on agriculture.

AIJ was conceived in 1995 at the first Conference of the Parties to the UNFCCC, and was intended as a 'learning by doing' exercise (Brown and Corbera, 2003, S43). Only 17 per cent of the activities were LULUCF-related, but one of the few AIJ that was relevant to agriculture was the

Scolel Té project in Chiapas, Mexico. Scolel Té is still very much in progress; it is not scheduled to finish until 2018 (Seeberg-Elverfeldt and Tapio-Biström 2010, pp12–13), and is accounting for carbon over 100 years (1997–2097). It has sold carbon to customers such as the International Automobile Federation, which up to 2004 had bought 5500 tonnes a year from the project at $11 per tonne; the project also sold carbon to the British 'retail' carbon trader Future Forests (now the Carbon Neutral Company) (Corbera, 2005, p45).

Scolel Té is fundamentally a forestry project (its name means 'growing trees' in Mayan), and is certified by the Rainforest Alliance. However, the activities have been taking place within the context of a subsistence farming system based on maize and beans, with some extra income from coffee, cattle and itinerant labour, and the majority of the carbon contracts have been with farmers who earned their living that way (Nelson and de Jong, 2003, p20).

Indeed, from 1998 some contracts did cover agricultural activities such as rotation of corn with a cover crop of mucuna (*Mucuna pruriens*). This appears to resemble the *abonera* system; used in Guatemala and Mexico, this has also spread, with migration, to Honduras (Laing and Ashby, 1993, pp62–64). *Abonera* replaces the traditional maize-cropping cycle with a single planting a year, intercropped with mucuna beans. An experiment in Rio Grande do Sul, Brazil found that a mucuna/maize rotation plot sequestered 15.5Mg/ha CO_2 over eight years, compared to a net emission of 4.32Mg/ha CO_2 from the traditional maize/fallow plot (Evers and Agostini, 2001, p6). The mucuna involvement evolved, according to Nelson and de Jong (2003, p26). 'In some regions,' they report, 'the carbon broker has been experimenting with contracts for various "carbon products" for possible sale under future global climate agreements such as a three-year cycle of corn/mucuna planting. The argument is that a subsistence corn system that does not burn residuals before the new planting cycle would reduce carbon emissions.' This is definitely an Article 3.4 activity.

Scolel Té has always been intended to be transparent and verifiable. Tipper (2002) says that the feasibility project, which began in 1994 with funding from the Mexican government and the United Kingdom, called for Scolel Té to be 'evidence-based' – meaning that: 'Overall quality and credibility of the system should be based on verifiable, documented evidence in the form of field data, accounting records, published literature, and official statistics.'

The activities to be monitored are defined by the farmers themselves, who – individually or in groups – approach the project trust fund, the *Fondo Bioclimatico*, with a proposal. The fund provides advice at the planning stage if necessary. At approval, the participant is issued with a passbook, or 'carbon bank book', in which the tonnage of carbon accumulated is registered, following a monitoring procedure which varies according to the broad type of activity undertaken.

The system has evolved as the project has grown. Between its inception in 1995, and 2004, the project grew from 6 communities consisting of 42 farmers to 33 communities with a total of 650 farmers. Initially, the technician from the local implementing NGO knew the participating farmers, and kept files on their contracts and monitoring procedures. When he left, however, much of this knowledge went with him, and the two technicians who replaced him in 1998 were forced to tighten the administration, eventually instituting the passbook system and keeping a duplicate of each book themselves. As Nelson and de Jong (2003, p26) record: 'It was only in 2001 that the technicians began to have confidence that their database reflected the reality of the carbon projects in various communities.'

There were further reforms. A monitoring system evolved by which the technicians review 10 per cent of the plots in each community. However, this has had its drawbacks. Technicians began to spend much more time on such administrative duties than they spent providing technical advice; not all the farmers were happy about this, but the technicians had no option as the project does not generate enough revenue for them to do both. Indeed Nelson and de Jong point out that the project evolved from sustainable development concerns to becoming a much more focused carbon 'bank' (Nelson and de Jong, 2003, pp27–28). There were also potential equity and biodiversity concerns about the project's evolution (Nelson and de Jong, 2003, p26; Brown and Corbera, 2003, S53–S54; Corbera, 2005, pp48–50; these areas are discussed more later in the book, as they will take a different form in a purely agricultural project).

For the moment, there seem to be two main lessons to draw from the Chiapas project: first, that a system of on-the-ground monitoring is possible, perhaps by using the 'bank book' model; but that second, this system will make demands of its own on personnel resources, which may thus be diverted from sustainable development objectives. These demands may be heavier in soil carbon projects as they will demand more specialized tools, especially if NIRS or flux measurements are to be used.

More recently, the GEF project *Integrated silvo-pastoral approaches to ecosystem management*, implemented in Colombia, Costa Rica and Nicaragua between 2001 and 2008, has raised some of the same questions. This project has been widely reported on elsewhere (see list of reports and publications in World Bank, 2008, pp63–71), so it will be described here only briefly.

The project cost a total of about $8.7 million, of which roughly half came from GEF. The purpose of the project was to test the PES paradigm in areas of degraded pasture (World Bank, 2008, piii). Such pasture is an environment that has considerable potential for carbon sequestration, especially in Latin America – the field study described later in this book is, to a great degree, concerned with this. However, the GEF project was specifically aimed at adoption of silvipastoral systems that integrate trees and livestock. The stated target was 12,000ha 'with improved biodiversity and carbon sequestration indices', a target that GEF claims was achieved and in fact slightly bettered (World Bank, 2008, piii).

It was also anticipated that about 4000ha would be converted to silpastoral systems, a target that was narrowly missed because, according to the final evaluation, 27 farms were sold during the project (World Bank, 2008, piv). The project also aimed at incremental carbon sequestration of 25,000 tonnes in the areas of PES, and reported that about 80 per cent of this had been achieved.

In addition, a 35 per cent reduction was reported in CH_4 and N_2O emissions due to improved fodder quality – although the report also says that stocking rates were improved from 1.8 to 2.5 cows per hectare, so it would be important to know whether the CH_4 and N_2O reductions were overall or per livestock unit. There was also a claimed reduction in soil erosion (although, as has been explained in this chapter, soil erosion is notoriously difficult to measure; though that does not mean the claim is not accurate). The report also points to an improvement in species diversity.

Finally, it is reported that farm income of participants was increased by 10 per cent during the project period. This is of some importance as the beneficiaries were small- to medium-scale farmers with between 10 and 80ha each, and an average annual income from the farm of around $3000 (op. cit., p3). So the PES payments per farm between 2003 and 2008 – which were US$2500, US$2400 and US$2300 for Costa Rica, Nicaragua and Colombia respectively (World Bank, 2008, p12) – would have had genuine impact.

However, GEF accounts the project as a success in two other important respects. First of all, although PES was needed to make the farmers adopt the practices, GEF reports that they proved profitable enough for farmers to maintain them without external incentives. This is crucial to the permanence of carbon sinks. Any practices proposed to farmers should be to their advantage, or the sink will not be maintained, with accounting implications that were described earlier.

Second, the project achieved what a simple market-based project cannot so easily do – it integrated other environmental goods, such as biodiversity and water quality, with carbon. It did this with an environmental services index – the Ecological Index – which allocated differing points for various practices and types of land use. Villanueva et al (2011) explain that the land uses were ranked from 0 to 1 according to their biodiversity value, and from 0 to 1 on their carbon storage; these two measurements were then aggregated. Thus primary forest scored two points, and degraded pasture none. In between, improved pasture in good condition, but without any trees, scored 0.5, while improved pasture with a high tree density scored 0.9. Payments were fixed accordingly.

Villanueva et al explain how payments were handled in Costa Rica. Land-use data was from satellite images but was verified with a farm visit; the data was then digitalized and the farm assigned the necessary value. During the first year of the scheme, farmers were paid $10 per point according to the baseline value; in subsequent years they would be paid up to $110 for each incremental point.

There were drawbacks. One was that those most benefited by the project initially were those with a low baseline – e.g. farmers who had (say) a large amount of degraded pasture to begin with. Villanueva et al also report that monitoring costs – which they calculated as an average of $4.18/ha – were less than half of this on farms of over 40ha. 'This result illustrates that monitoring costs may be a barrier for the participation of small producers,' they say, and suggest that numbers of smaller farms might be organized as single units for monitoring purposes. Moreover the authors reported wide variations in monitoring cost according to land use.

Villanueva et al also broke down the monitoring costs and found that preparation of field equipment, and collecting and processing information using geographical information systems, accounted for 46 per cent of the cost; farm visits and monitoring of land uses accounted for 31 per cent; equipment and materials 16 per cent; and transportation 7 per cent. So, despite taking a more high-tech approach than Scolel Té, and despite the innovation of the Ecological Index, the three-country project faced the same difficulties with transaction costs. Nonetheless, Villanueva et al report: 'As a tool for monitoring and for payment for ecosystem services, the ecological index method costs less than current quantification techniques. Moreover, it is easy to understand for the service providers and it is flexible.'

Whether this approach would fulfil the requirements of a project generating CERs for the compliance markets is less clear; but it was designed to reflect other environmental goods as well. On the evidence, it would seem that PES projects of this type can sequester carbon. The arrival of REDD+ may increase the number of such PES projects, and is discussed next.

REDD and REDD+

Avoided deforestation was not included in the CDM at the Marrakesh negotiations because of concerns about leakage, and because the amounts of credits involved could be so large as to flood the market (Schlamadinger et al, 2007b, p278). However, at COP13 in Bali in 2007, it was agreed that reduced emissions from deforestation and degradation (REDD) would be part of any Kyoto II after 2012.

There has also always been opposition to REDD because (as with PES in general) it pays people to do what they should be doing anyway. However, there is an overwhelming economic case. During the 1990s, emissions from deforestation were 5.8 gigatons (Gt) a year, a fifth of the total anthropogenic carbon emissions and more than those from the entire transport sector (Karousakis, 2007, p6).

Besides, REDD looks like good value against other emissions reductions, including afforestation and reforestation. Swallow et al (2007, ppvi–vii) report that in three areas studied in Indonesia where emissions were increasing, the economic returns from deforestation were less than US$1 per tonne of carbon emitted in up to a fifth of cases, while in others in Indonesia and Peru, emissions from land-use change generally returned less than US$5 per tonne. The price of carbon fluctuates, but would normally exceed that by a wide margin. Chomitz et al (2007, ppxii, 1) argue that since clearance of Latin American tropical forest to create pasture releases 500 tonnes of CO_2 per hectare to create pasture worth only a few hundred dollars, monetizing the carbon should do the trick. 'This implies … that deforesters are destroying a carbon storage asset theoretically worth $1,500–10,000 [for carbon trading] to create a pasture worth $200–500 (per hectare),' he says. This is also a powerful argument for pricing reduced deforestation. Since 2005, REDD has evolved into REDD+, applying to reduction on forest degradation as well as deforestation: put crudely, REDD+ is about forest quality as well as area.

Including avoided deforestation in world climate governance makes the inclusion of agricultural sinks nearly inevitable. REDD+ does not in theory cover sequestration of carbon into soil sinks (Pirard and Treyer, 2010, p5). But as Bloomfield and Pearson (2000, p13) point out: 'Only half of the land that is converted from tropical forest to agriculture actually increases the productive agricultural area; the other half replaces previously cultivated land that has been degraded and abandoned from production. … Thus, if existing agricultural land can be used more effectively, less forest and savanna land will need to be converted to cropland and pasture, reducing the greenhouse gas emissions associated with land conversion.' Evidence for this has been presented on a global scale by Vlek et al (2004); at local level, Tschakert and Tappan (2004) noted the relationship between intensification and extensification in a carbon-related study in Senegal. This implies that building up soil organic matter on the farm will help protect forests.

But the relationship between agriculture, intensification and extensification, and deforestation is a difficult one. It is of course logical that the productivity and profitability of farming a given stretch of land will have a bearing on changes in land use, and indeed this relationship was quantified by the Mecklenburg landowner and economist Johann von Thünen as early as 1826 (Angelsen, 2009, p128; Rudel, 2009, p194). According to Angelsen, this has been borne out by more modern research. 'A survey of more than 140 economic models of deforestation finds a broad consensus on three immediate causes of deforestation; higher agricultural prices, more and better roads and low wages coupled with a shortage of off-farm employment opportunities,' he states (Angelsen, 2009, p128). He suggests a number of policies that might reduce deforestation for agricultural purposes, including depression of agricultural prices and creation of off-farm opportunities. Angelsen regards these as high-efficiency in reducing deforestation, but the field survey in Brazil described later in this book suggests that both those policies could cause a loss of carbon from existing farms, accompanied by loss of productivity that could in itself drive extensification.

However, the third option he quotes – support to agricultural intensification – might have more positive results in not only reducing deforestation but increasing the carbon pool on existing farms. It is this that makes avoided deforestation a potential lever for both carbon sequestration and poverty alleviation in agriculture.

There are some assumptions here, not least that forest clearance is first and foremost for agriculture; this is generally true, but the causes of tropical deforestation can be complex. Moreover, the interaction between agriculture and deforestation is not always predictable. For example Carpentier et al (2000, p75) suggest that increasing the profitability per hectare of farmland will increase the incentive to deforest. Pirard and Treyer (2010, p7) argue persuasively that the 'intensification reduces deforestation' argument is flawed, because it assumes fixed demand for food. But consumption of meat, for example, is rising, and that will increase demand for pasture, which will in turn increase supply relative to price and increase demand. 'Although intensification – in other words the increase in productivity per hectare – is a key variable for long-term forest conservation, the problem cannot be resolved by this alone,' say Pirard and Treyer. 'The scientific findings ... indicate that there is no simple, unequivocal relationship between changes in agricultural systems and tropical deforestation' (Pirard and Treyer, 2010, p13).

Pirard and Treyer do not oversimplify this debate; neither does Rudel. Rather, they suggest common-sense measures to guard against perverse outcomes. Pirard and Treyer (2010, p16) suggest that assistance to intensification needs to be combined with PES-type measures. Rudel (2009, p199) also advocates using both, but in separate locations, with intensification being applied in agriculture close to urban centres and well away from the forest frontier.

But will this appreciation of complexity find its way into policymaking, especially at local and regional levels? There is a nightmare scenario in which farmers, helped to intensify, see their profits per hectare rise, employ more labour and decide to expand, cutting down every tree in sight. Given the huge sums likely to be involved in REDD+ – perhaps $2–10 billion in the early phases (Sunderlin and Atmadja, 2009, p50) – there is potential for a mighty train wreck.

One safeguard might be to incorporate REDD+ into unitary national land-use policies, so that policymakers have a clear overview. In fact, there was a move to integrate non-forest land use with REDD+ in a single sector, known as AFOLU (Agriculture, Forestry and Land Use), but this was rejected at COP14 in Poznan (Loisel, 2008, p2). Carbon stocks in agricultural and forested land can of course still be accounted for together at national level. But the resolution reached on REDD+ at Cancun does allow for sub-national accounting, at least in the interim while national mechanisms are strengthened.

However, REDD+ may not turn out to be the juggernaut that it appears. Although much donor money may be involved, the 2010 Cancun agreement stopped short of the use of market mechanisms as well for REDD+, as a result of strong opposition from Bolivia (some NGOs also have strong views on this). This may change; but if it does not, it may reduce the significance of REDD+, especially given that some early finance has been private capital speculating on generation of compliance credits (Sills et al, 2009, p268). Some of this was predicated on US legislation that, as explained in the previous chapter, has not emerged, at least at the federal level.

That does not mean that REDD+ will be a failure, but it may mean that it will mostly take the form of public-sector PES schemes of the sort described above. Even before Cancun, Seymour and Angelsen (2009, p299) suggested that: 'In the medium to long term, PES schemes are likely to be the implementation instrument of choice.' REDD+ may therefore be evolutionary rather than revolutionary. Nonetheless, if Rudel's suggestion – to improve agriculture away from the forest frontier – is adopted, it could spark a broader range of activities in agriculture than existing PES schemes have done.

The Practical Questions

This chapter has aimed to answer practical questions that would arise in the inception and design of agricultural sinks projects. It did not consider livelihoods concerns, as these are reviewed in the discussion of the case-study survey methodology in the next chapter.

The chapter began by asking whether anyone would invest in agricultural sinks. In the developing world, perhaps not; the abatement costs per tonne would be too high. However, an alternative model would be for project developers to meet the up-front costs themselves, and sell the credits afterwards – what is known as a 'unilateral' project.

Next, the chapter looked at baselines and additionality; there are problems setting the first, and ensuring that carbon content is additional to it. This is especially difficult with agricultural sinks because the soil carbon content was probably not in equilibrium at the time from which the baseline must be set – even assuming one knows what the SOC content then was. The only answer to this may be to have control plots so that the baseline could be monitored during implementation.

A further challenge discussed was leakage. A major source of this might be through extra use of inputs that had themselves generated emissions in their manufacture, transport and use. However, these emissions can be accounted for. More difficult would be leakage from systemic changes due to new management practices, especially regarding the use of crop residues. The only answer to this is due diligence during project assessment.

This was followed by a discussion on permanence. This is a difficult area for land-use projects, but there is no reason why that should be any more true of agricultural projects than forestry. It is really an accounting problem. The solution suggested here was a form of tonne-year accounting in which the carbon is effectively leased, with payments reflecting the declining effect of each tonne on climate change. However, even where this is to be done, there should be a reasonable expectation of permanence, or there is a risk sinks will do more harm than good.

Monitoring and verification has been a major constraint to the adoption of agricultural sinks under the Kyoto Protocol. There is no fundamental problem with measuring soil carbon, but it is subject to error due to spatial variability. It will therefore be expensive if a high confidence level is required. However, new tools have appeared in recent years and one of the most promising, NIRS, may eventually permit much cheaper *in situ* assessment. In the meantime, an activity-based system of stratified accounting was proposed, starting with traditional soil sampling at project inception, followed by the use of Landsat data to verify implemention of agreed management practices. Ground-truthing will be necessary.

This chapter has also reviewed PES schemes as perhaps a simpler and practical alternative, and there was a brief discussion of the long-established Scolel Té project and of the three-country GEF-assisted project in Latin America. These projects have shown what can be done where public finance is available and the standards of verification required to generate formal CERs are not needed. Should it be decided that accurate quantification of soil carbon is just too difficult and CERs cannot be generated, this is the alternative. The donor finance available under REDD+ may help.

However, the overall conclusion is that there *are* no insuperable constraints to generation of CERs. Marketing of agricultural sinks, baselines and additionality, leakage, permanence, monitoring and verification, and project administration all present very real challenges, but they can be done. If a single major constraint emerges from this chapter, it is whether or not agricultural sinks in poorer countries will be economic. That depends, at least in part, on the future price of carbon; and that nobody can foresee.

Notes

1 The United Kingdom, Germany, The Netherlands, Australia and the United States. Jung et al took the figures from a 2005 presentation to the UNFCCC by Surya P. Sethi, energy adviser to India's Planning Commission, who gives the Tata Energy Research Institute (TERI) as their original source. The lower figure includes China as well as India.
2 Soil Organic Carbon Reserves and Transformations in Agro-Ecosystems.
3 There may be a problem after the first 20 years, as most soils would accumulate carbon only slowly after that point. Again, however, global per-capita carbon allowances would provide an incentive for host-country governments to address this. A more serious problem might be something like a sudden demand for cereals, giving the farmer an incentive to farm unsustainably – as in the example of the Soviet grain purchases cited by Blaikie (1985).
4 The ^{137}Cs technique works by detecting the presence or otherwise of isotopes released during the nuclear tests of the cold war era; simply put, they will be present if no soil has been carried away, but it if has, they will go with it. However, there are constraints to large-scale uses of this practice; Stroosnijder (2005, pp169–170) mentions a number, among them the need to differentiate from isotopes released by the Chernobyl disaster in 1986, the cost of analysis and the need for undisturbed control plots such as graveyards, which may or may not be available and genuinely undisturbed.
5 There is more than one definition of biodiversity but Article 2 of the United Nations Convention on Biological Diversity states: '"Biological diversity" means the variability among living organisms from all sources including, inter alia, terrestrial, marine and other aquatic ecosystems and the ecological complexes of which they are part; this includes diversity within species, between species and of ecosystems.'

6

From Theory to Practice: The Atlantic Forest Biome

Nature has done everything to help us, but we have done nothing to help Nature. ...Our precious forests are disappearing, victimized by fire and by the destructive axe of ignorance and selfishness. Our mountains and hillsides are daily balding, and in time there will be a shortage of the fertile rains that appease the vegetation and feed our watersheds and rivers, without which our beautiful land of Brazil will be reduced, in less than two centuries, to the condition of the empty plains and the arid deserts of Libya. (José Bonifácio de Andrada e Silva (1763–1838), 1823; quoted in Pádua, 2004, p4)

The preceding chapters have reviewed the scientific, political and economic bases for carbon sequestration in agriculture. So far, however, the farmers themselves have been absent, except as participants in a mechanical process by which money would be given and carbon credits received. It will of course not be like that, as such on-farm initiatives must be reconciled with farmers' priorities in the context of the broader economy. It may be that some carbon sequestration practices are unattractive to farmers for practical reasons that are not obvious to the outsider. It may also be that some practices can be implemented locally, whereas others require policy changes or even macroeconomic shifts at a higher level. The methodology used in the case study described in the following chapters has been informed by the literature on technology transfer, which suggests that trying to influence farmers' behaviour can be difficult and is sometimes a bad idea.

The case study is based on the Atlantic Forest biome[1] of Brazil. The biome is one that has, historically, lost much of its terrestrial sink, but has considerable theoretical potential to sequester it back. Doing so may also serve other environmental goals; they include biodiversity and forest protection – but also reduction of pollution from agrotoxins, a problem of which the farmers are often well aware. The methodology chosen reviews proposed carbon-friendly practices within the context of farmers' own perceptions. It covers a wide range of management practices – a 'basket of technologies' approach that was suggested by prior experience on technology adoption and resource use. This is explained later.

This chapter begins by describing how the Atlantic Forest biome has assumed its current form, essential context for this case study. It then reviews the carbon sequestration potential of the biome. This is followed by more specific information on the fieldwork locations. In the following chapter, the methodology is discussed with reference to the literature that influenced its design. The actual results are presented and discussed in chapter 8.

The Atlantic Forest

On Easter Sunday, 1520, the Portuguese explorer Pedro Álvarez Cabral made landfall near present-day Bahia in north-east Brazil. For the following six weeks, his expedition sailed south along the coast, and Europeans saw the extraordinary stretch of largely unbroken forest that is today known as the Mata Atlântica – the Atlantic Forest – for the first time.[2]

It is not clear exactly how large it was. In 1993 the Instituto Brasileiro de Geografia e Estatística (IBGE) estimated its extent in 1500 at 1,363,000km^2, covering 16 per cent of what is now Brazil and spreading across 17 of its modern states; but some estimates have put it rather higher (Câmara, 2003, p35). Even the IBGE estimate would put it at roughly 25 per cent of the size of the Amazon basin itself. But there is little left today. In 1995 it was estimated at 98,878km^2, or 7.25 per cent of the extent in 1500 as estimated by the IBGE (Câmara, 2003, pp35–36).

The remaining forest may lack the importance of the Amazon in terms of above-ground carbon, but at nearly 100,000km^2 it is still substantial. Moreover it is an area of very high biodiversity, and this can also be used to leverage funding for carbon maintenance; in fact the two are synergistic goals as maintenance of soil fertility can, at least in theory, slow down the loss of the forest that contains that biodiversity. This is in effect what is happening through the GEF-assisted Rio Rural project in Rio de Janeiro State, discussed later.

So carbon aspects should not be considered in isolation from the forest's biodiversity status. Although some species were shared with the Amazon,

Table 6.1 *Species, threatened species and endemism in Brazil's Atlantic Forest*

	No. of species	Endemic species	Endemism (%)	Threatened species (no.)	Threatened species (%)
Trees and shrubs	~20,000	~8000	40	367	2
Birds	849	188	22	104	12
Mammals	250	55	22	35	14
Reptiles	197	60	30	3	1.5
Amphibians	340	90	26	1	0.3

Source: adapted from Tabarelli et al (2003)

the original level of endemism was very high, and it is still spectacular. The late Warren Dean, who wrote the standard work in English on the destruction of the forest – *With Broadax and Firebrand: The Destruction of the Brazilian Atlantic Forest*, cited hereafter as Dean (1997) – reports that 270 tree species have been found in a single hectare in South Bahia and that while 8 per cent of the tree species were shared with the Amazon, more than half were endemic (Dean, 1997, p14). The forest remains an important biodiversity hotspot, as Table 6.1 shows. Of the 20,000-odd tree and shrub species, some 8000 are endemic (Tabarelli et al, 2003). Neither is this diversity restricted to plant species, with the biome still a refuge for a wide range of birds and animals. These include globally endangered animal species such as the maned three-toed sloth, the woolly spider monkey, the red-browed Amazon parrot, the black-headed berryeater and the solitary tinamou. Two of the endangered species, the plumbeous antvireo and the buffy tufted-ear marmoset, are at least partly dependent on the area of the Noroeste Fluminense (GEF, 2003a, p4), in which one of the case-study sites was located.

Despite this, deforestation continues apace. The NGO that fights for the forest's preservation, the São Paulo-based Fundação Mata Atlântica, claims that over 10,000km^2 was cut between 1985 and 1995, and that a further 2.84km^2 go every day (GEF, 2003a, p37). Hirota (2003, pp61–62) says that the State of Rio de Janeiro alone lost over 300km^2 in the second half of the 1990s. In May 2005, the author travelled along BR116 from Rio de Janeiro towards São Paulo with an expatriate who had not made the journey for about 20 years; the latter was astonished at how much of the forest that used to flank the road had gone, replaced apparently by rough pasture for cattle.

Legal instruments exist to protect the forest, but can have unintended consequences. For example, as Tabarelli et al (2005, p696) point out, the 1965 Forest Code requires amongst other things that 20 per cent of any rural property in the biome should be managed as a legal forest, known as *reserva legal*; the farmers surveyed for the case study were presumably aware of this, as they often referred to forest as *reserva* rather than by the Portuguese word for forest, *mata*. However, as Evans (2004, p32) reports, 'It was possible [until 1993] to clear 80 per cent of a property, then *sell* the legal reserve. The buyer could then clear 80 per cent of his new land, leaving a new legal reserve, *which could also be sold* [italics in original].'

The tax regime has also had unintended consequences, discriminating as it does in favour of land in agricultural use, rather than *reserva* (Tabarelli et al, 2003). This has had a further effect in Rio de Janeiro and São Paulo states; the author was told by Brazilian colleagues, and by extension staff, that this encourages those holding land as capital to farm it, but farm it badly, as it is not their prime purpose for holding the land; indeed they are often doctors or lawyers from the two large cities. It is therefore supervised by poorly skilled staff and no investments are made in (for example) pasture improvement; it becomes not merely deforested but degraded.[3]

In the meantime, economic pressures to cut the remaining forest over the last 30 years have included the oil crisis in the 1970s, which led to the cutting of much forest in São Paulo to grow sugar cane for alcohol production (Câmara, 2003, p37); this is a serious drawback of biofuels as a climate-change mitigation strategy (although changes to the law on trash-burning mean that sugar cane is probably not driving deforestation now). More forest was replaced at this time with imported fast-growing species for cellulose and paper production (Câmara, 2003, p37).

There have also been political pressures, and there is still a permanent lobby that favours expansion of agriculture and other land uses (Tabarelli et al, 2005, p696). Brazil does care about its environment and natural resources, but its people do not always accept the preservation of forests as an automatic good and indeed are suspicious of foreign environmentalists' intentions. As a report to an agribusiness centre in the US Midwest has put it: 'It is ... conjectured that international groups lobby for the preservation of natural forests to prevent Brazilian farmers from expanding production. ... [Some] feel Brazil's agricultural potential is being thwarted by farm subsidies in the United States and international trade agreements that favor the developed nations' (Matthey et al, 2004, p19).

However, in order to understand how carbon stocks (both in forests, and in agricultural soil) can be restored, it is necessary to understand not just these recent threats to the remaining forest, but the process that destroyed the 93 per cent that has gone. How did the Portuguese colonists and their successors manage to destroy a forest 12 times the size of their own country? What did they replace it with, and why? The answers to these questions explain much about current patterns of extensive rough pasture and sub-optimal management of organic matter. As Dean (1997, p5) puts it: 'South America … is the forest historian's freshest battle-ground, where all the fallen still lie sprawled and unburied and where the victors still wander about, looting and burning the train.'

The Decision to Despoil, 1500–1823

When the first Portuguese settlers arrived, there was already nothing new about the clearance of the forest for agriculture. True, the early inhabitants seem to have lived mostly on the seashore and to have subsisted on a diet of easily gathered shellfish; but long before 1500 they had been displaced by the Tupi people from farther south, who did grow crops. Dean (1997, p25) reports evidence of maize in Minas Gerais as early as 1900 BC; in fact agriculture had been practised elsewhere in South America for some time, but Dean believes that the forest abundance had made it unnecessary in the Atlantic Forest. He thinks that population increase, rather than climate change, eventually changed this.

The Tupi practised a system of shifting cultivation using cassava, which would take about 18 months to reach maturity; the land would be cleared with fire, used for two or three such seasons and then abandoned, either because the soil was exhausted or, just as likely, because of invasive weeds or pests or because the settlement needed to move (Dean, 1997, pp26–29). The resulting forest regrowth, distinct from primary forest, was called *capoeira* by the Portuguese. Dean argues that, on the basis that the Tupi cleared only primary forest, and given an estimated population of six to nine million in 1500, about half the forest would have been burned and regrown in the 1000 years before the Portuguese arrived, even assuming none was burned for other reasons – so what they saw was often not virgin forest at all. They seem to have been aware of this, and the earliest surviving land grants in Rio de Janeiro, dating from the 1590s, generally refer not to *matos verdadeiros* (virgin forest) but to *matos maninhos*, which

Dean (1997, p35) thinks referred to secondary regrowth. He also argues that, left alone, the Tupi might themselves eventually have exhausted the resource base and gone into decline (Dean, 1997, p39).

In the event, the forest was to be dealt with much more quickly. The settlers adopted the Tupi method of slash-and-burn, but introduced new crops: fruit trees, cereals – and sugar cane; this did better than it had on Madeira and São Tomé, where it had needed manure. As early as 1535, the Portuguese also introduced cattle. Pastured in grassland around Guanabara Bay, where the city of Rio stands today, they faced few pests and diseases or predators (although, according to Dean, jaguars eventually took an interest in them). They also had ample pasture (Dean, 1997, p75).

The cattle did not, however, mix well with unfenced slash-and-burn cropping, and when the existing pasture was exhausted, forest clearance seems to have been extensively undertaken for their benefit, leading to the beginnings of the extensive pasture system that can still be seen in the region today. Before long, land was being exhausted and abandoned, and grants for replacement land being made, reflecting an improvident land use that, says Dean, was 'utterly beyond the conception of the miserable Portuguese peasantry, whose households might be obliged to make a living from a single hectare yet be expected to pass it on from generation to generation, its productivity undiminished' (Dean, 1997, p77).

Dean may be right to ascribe the wasteful use of land to improvidence, but one can speculate that the separation between cattle and crops, besides driving forest clearance, also prevented the development of sophisticated integrated crop/livestock systems of the type one might have found in southern Europe or the Middle East, where organic matter is exploited well by moving it between the two halves of the system. However, Dean also identifies another driver behind the wasteful use of land: slaves. 'Not only were the short-lived slaves only briefly attached to the soil, they were, whenever possible, strangers to it,' he says, adding that in any case, the conservation of natural resources was to prove irrelevant in a society in which the conservation of human life was irrelevant (Dean, 1997, p56).

In fact, by the end of the 18th century, a small but growing class of thinkers in the colony had begun to develop a theory of natural-resource use that foreshadowed Dean's view. A Mineiro (Minas Gerais) magistrate, José Gregório de Moraes Navarro, wrote in 1799 that the farmers were reducing the trees to ashes, farming crudely and depriving the land of its resources, then moving on, leaving it exhausted and covered in invasive weeds (Pádua, 1999, p3). Introduction of a more intensive agriculture, he argued, would ease the burden of the slave. For Navarro, one of the keys was the plough – ironically, given that no-till agriculture has been developed for sustainability in modern Brazil; but it is not hard to follow his logic, for the plough symbolized for him intensification over extensification.

Others also questioned the slave-driven, land-hungry manner of farming in the colony. Brazilian writer and politican José Bonifácio de Andrada e Silva wrote in 1825 that if landlords had not had huge numbers of slaves at their disposal, they would not have cut down vast swathes of forest but would have re-used already-cleared land that now lay derelict. '[I]f agriculture is practised by the free arms of small proprietors, or by journeymen, these neglected lands will be used … and in this manner the ancient virgin forests … will be conserved as a sacred heritage for our posterity,' he wrote (Silva, 1825, quoted in Pádua, 2004, pp25–26).

There was in fact alarm at the way in which the country's natural resources were treated. Pádua (1999, p4) states that between 1786 and 1888, at least 38 Brazilian authors wrote on the environmental challenges of deforestation, exhausted soils, disturbance of the climate and the social problems caused by environmental degradation. The techniques adopted for cultivation and processing of sugar cane received attention; for example Manuel Arruda da Câmara wrote in 1799 that cultivation of this crop, which required 'the most profound knowledge of physics and chemistry', had instead been entrusted to ignorant and lazy men whose use of fuel was appallingly wasteful (Pádua, 1999, p4).

This tide of intellectual enquiry into Brazil's relationship with its environment seems to have arisen from a reinvigorated administrative and intellectual life in Portugal, inspired respectively by the distinguished reformist minister the Marquês de Pombal and the Italian scientist Domenico Vandelli; the latter's leadership at the University of Coimbra in the late 18th century brought an enlightenment spirit that was taken back to Brazil by many who studied there.[4] Indeed Pádua (ibid.) is surprised that this phenomenon is so little known not only internationally but also in Brazil. That this atmosphere did not prevail in Brazil is the more surprising in that, following independence in 1822, José Bonifácio de Andrada e Silva became Minister of the Kingdom (effectively, Prime Minister) and Foreign Affairs.

But he did not long retain high office. He advocated protecting the environment by abandoning the colonial system of large-scale, extensive, wasteful farms; moreover he also supported emancipation of slaves, albeit over time (Pádua, 2004, p23). Such views were not welcome to the new country's ruling elite and Bonifácio was dismissed and exiled. He was to return to Brazil but would die there in 1838, according to Pádua a broken man. For Brazil's forest and countryside – and its carbon stocks, below and above ground – the die was cast. The way in which agriculture affected the landscape over the next 150 years reflects extractivism and the wasteful agriculture that accompanied it. The change in the landscape would be driven, at least in part, by coffee.

Coffee

Coffee can be cultivated sustainably, but historically in Brazil it was not. It drove much of the deforestation process and, by its wasteful nature, provided a bridge between forest and the rough pasture often found in the Atlantic Forest region.

Coffee seeds may have first been brought to Brazil as early as the late 1600s, but really arrived after an army officer appropriated some seeds while on a mission in French Guiana in 1727. Thereafter it was grown in a small way, not least as an ornamental plant; still, Captain Cook found that Rio was still importing coffee from Lisbon when he passed through in 1768. By 1790 only just over a tonne was being produced for the local market.

In the 19th century, however, it was found that the hills of Rio de Janeiro State were well-suited to the crop, which needs plenty of rainfall – about 1300–1800mm a year – but does not tolerate waterlogging (Boddey et al, 2003, pp601–602). This discovery coincided with an astonishing explosion in demand. Topik (1987, p59) records that it increased by 25 times in the 19th century, driven by the growing population of the United States and the increasing prosperity of the working classes there and in Europe. By the middle of the century, Brazil was producing 70 per cent of the world's coffee.

As cultivation expanded up the Paraíba Valley in the north of Rio State in the middle of the century, in the dry season 'a yellowish pall hung over the province ... obscuring the sun by day and obliterating the stars by night' (Dean 1997, pp184–185). Dean records that some of those clearing the forest experienced feelings that were 'weirdly pyrophiliac'. Worse, Brazilians did not initially realize that coffee could be replanted, and thought it required virgin forest to be cleared. Thus in 1862 the Baroneza do Paty could write of the enormous holdings she had just inherited that: 'The absolute shortage of land for planting coffee did not allow me to increase any plantings. … Regretfully I must report that on all our *fazendas* which cover an area equal to 21,104,000 square *braças*[5] or almost two and one-half square legoas … we do not have 200 square *braças* of virgin forest of first quality' (Stein, 1957, pp45–46). The assumption that one must clear virgin land after a cycle of coffee (perhaps just 15–25 years) was not to be challenged until, in the Paraíba Valley at least, the damage had been done. Dean (1997, p186) comments that chemical analysis of coffee's requirements was not attempted until the 1870s, and was not 'connected to practice until much later'.

Moreover the early planters appear to have been ignorant of the conditions under which coffee was grown elsewhere, and did not grow it under shade. Soil erosion was also inherent to the techniques used. Coffee was planted in wide rows of 4–5m and the vegetation between was cleared away, even on steep slopes; slaves with heavy hoes worked in gangs down the slope for ease of supervision (Dean, 1997, p186) and harvested carelessly. Stein's 1957 monograph on Vassouras, a municipality in the Paraíba Valley that saw the first great expansion of coffee, describes how it saw a 'complete economic cycle which started with tropical forest and terminated with denuded, eroded slopes' in just a century (Stein, 1957, p289). Dean (1997, p268) states that coffee groves in northern Paraná possessed 165 tonnes of humus per hectare when planted but that this declined to 10 tonnes after just 20 years. It is not clear when these figures were collected (or exactly what Dean understands by humus), but extensive clearance for coffee has continued in that area within the last half-century. This suggests catastrophic loss of soil carbon as well as productive capacity.

For the planters of Vassouras, the *coup de grâce* was administered by the end of slavery in 1888. Although some slaves were persuaded to return to the plantations as free labour, at least for a short time, their capital value had represented the only collateral available to finance the exhausted and failing *fazendas*. However, coffee moved on, into Minas Gerais, São Paulo, Espírito Santo and into other parts of Rio de Janeiro. Between 1900 and 1930, global demand for coffee doubled (Topik, 1987, p59). By the late 1920s, coffee accounted for 71.1 per cent of Brazil's exports by value (Fausto, 1999, p234–235). The environmental effects of coffee production appear to have bothered few people; when an area was exhausted, the crop simply went elsewhere.

What did slow this process in the end was overproduction, encouraged by the government's policy of price protection. A record crop in 1929 coincided with the Depression and prices tumbled from 22 cents a pound in the late 1920s to 11 in the early 1930s and just 9 thereafter. The Brazilian government would be forced to destroy 57 million sacks of excess coffee (Topik, 1987, p84). As previous policies of price support had encouraged planters to take out large loans in order to plant more, many found themselves, as Fausto puts it (1999, p191), 'up a blind alley'. Coffee *fazendas* collapsed.

The story of coffee in Brazil did not end there, of course. Despite further price shocks in the 1950s, it has remained an important industry for Brazil; large areas of forest in northern Paraná have been cleared for coffee in the last 50 years, and there has been recent expansion into the Cerrado, the vast unexploited agricultural frontier between the Atlantic Forest biome and the Amazon (Boddey et al, 2003, pp601–602). Brazil is still the world's largest coffee producer, coffee is still its sixth largest planted crop, and some 1.96 million ha is still grown within the Atlantic Forest biome itself (Boddey et al, 2006, p317). But coffee is no longer grown on any scale in the case-study areas. What matters is what succeeded it, as it explains the continuing preponderance of extensive land use, the loss of carbon and the potential to get it back.

After Coffee, Cattle

There was never much chance that exhausted plantations would revert to forest, even were the soil still to support it. In the Paraíba valley, migrants from Minas arrived with cattle, which fed upon grasses that spread quickly through the abandoned plantations (Stein, 1957, pp286–288). The nature of these grasses is discussed in more detail later, but for the moment it should be noted that much of it was of poor quality as forage (for example, *colonião* species thought by some farmers today to be native, but probably introduced in the bedding of slave ships). Dean (1997, pp112–114) points out that 'compared to the varied native grasslands, pastures filled with a single exotic grass did not provide a balanced regime of amino acids and micronutrients'.

The invasion of exhausted coffee plantations by cattle spread through and beyond the Paraíba valley and became a key factor in the development of the biome. One reason for this may have been the decline of the once-abundant native pasture. As we have seen, cattle had arrived as early as the 1530s. Excellent early results quickly gave way to overgrazing of the most palatable species, which would be replaced by noxious weeds. Ranchers reacted with frequent burning; this would bring about a brief improvement but then altered the species composition, favouring tussocky growth that encouraged soil erosion (Dean, 1997, pp112–114). Increased soil compaction also resulted, favouring plants with short roots. 'The deterioration of native pasture caused it to become, paradoxically, a scarce commodity,' says Dean. It is this process that largely formed the Paraíba watershed as it is seen today, stretching through São Paulo and Rio de Janeiro states and into the Zona da Mata of Minas Gerais, and taking in the case-study sites.

However, not only was the landscape degraded, deforested and eroded; the people went too. In Vassouras, where the coming of cattle coincided with emancipation, there were fewer jobs: 'The cattle moved in and the people of Vassouras … moved out' (Stein, 1957, pp286–288). This process continued into our own time. Dean implies that pressure for land reform in the 1950s and 1960s encouraged landowners, who anyway saw cattle-raising as a more 'noble' profession than arable farming, to go over to pasture, and suggests that the labour laws of the military government that took power in 1964 assisted this. Farmers surveyed in the case study saw labour shortage as a major constraint to some of the carbon-friendly options put to them. True, rural depopulation is a deceptive phenomenon; many people who have moved out of agriculture have not gone far. This is discussed later. But the basic process by which the region has been shaped has been clearance for coffee, exhaustion by coffee, and its replacement by ill-managed pasture – leaving a landscape that is degraded and low in organic matter and soil carbon, and apparently also depopulated.

Can agriculture be used to restore it, and mitigate climate change at the same time? If not, why not? It was hoped the farmer survey would uncover the constraints to more carbon-friendly farming in the region. In so doing it may reveal whether the historical background has bequeathed a structural bias towards low-intensity, extensive systems that do not enhance sustainability. Breaking this cycle might be worthwhile if the sink can be substantially enhanced, so the next section looks at what the potential might be.

Potential for Carbon Sequestration in the Biome

There is of course a huge gap between the theoretical sequestration potential of a biome, and what is feasible in practice. Indeed Boddey et al (2006, pp337–341) go so far as to provide two completely different estimates, the higher based on technical feasibility and assuming maximum political will, the lower based on what they think might be feasible given probable policy changes in Brazil. The discussion in this section is based on this second scenario. This makes a number of assumptions based on the authors' political and economic predictions.

Some of these assumptions are given in Table 6.2, but they are in turn based on others – for example, the 40 per cent figure for pasture rehabilitation is based on continuing growth in demand for Brazilian beef (Boddey et al, 2006, p335). The authors have also assumed increases in demand for certain Brazilian crops as a response to decreasing developed-country subsidies following the Doha round of trade talks in 2003 (Boddey et al, 2006, p336). They have also made assumptions about changes in agronomic practices in sugar cane, following recent legislation on trash burning (Boddey et al, 2006, p338). Some of these assumptions are quite large (on subsidies, for example) but the authors are familiar with both the country and the region, and their assumptions are likely to be reasonable. In any case, the alternative is simply to discuss the ultimate biophysical potential of the biome, and in the real world that makes little sense.

The 'realistic' scenario suggests increase of the biome's soil carbon sink by 363Tg (363 million tonnes, or 0.363Pg) over a 20-year period. (The higher estimate is 790Tg.) Taking the price of carbon as \$3.50 a tonne – a very low estimate, but probably fair given the accounting and certification problems with agricultural sinks – and multiplying by 3.66 to convert to CO_2e, even 363Tg is worth US\$4684 million or over US\$230 million a year, more than enough to fund the sorts of measures that could make it a reality. Moreover, these figures are in some respects conservative; for example, the authors state that these figures take no account of above-ground biomass. But in tropical systems the above-ground biomass may be very high; sugar cane is a case in point. Biofuels offsets (including but not only those from sugar cane) could also have been discussed.

Table 6.2 shows, in adapted form, how Boddey et al divide the potential. Some are qualified; for example, the entry 'Rotations of annual crops'

Table 6.2 *Land-use change and soil carbon stocks in the Atlantic Forest biome under a realistic land-use change scenario*

	Area (million ha)	***Change in SOC, t/ha, 2005–2025***	***Total change for Atlantic Forest biome (Tg)***
Recovery of 10% of 'natural' pasture using leguminous tree species	2	25	50
Recovery of 40% of planted pastures using fertilizers, with or without tillage*	8.4	15	126
Rotations of annual crops	2	0	0
As above, using green-manure legumes with no-till	7.9	20	158
Sugar cane in place of degraded pasture	4	0	0
Increasing area of sugar cane in which trash conservation is used	5.3	5	26.4
Planted forest in place of degraded pasture	3	0	0
Converting 20% of coffee area to organic production	0.4	10	4
Cacao area decreased by 20%	0.12	−10	−1.2

*It is not clear to what extent reduced tillage regimes would assist carbon sequestration in pasture recovery. This is discussed further in the next chapter, as are rotations.
Source: adapted from Boddey et al (2006, p337)

shows no gain in carbon through rotations unless a legume is included in the rotation. The authors are reflecting the experience of Sisti et al (2004), who report that rotations with soybean and no green-manure legume sequestered no carbon. Boddey et al (2006, p336) assume that growing demand will result in some degraded pasture being planted to soybean in the biome and are arguing that the correct rotation would result in this having a positive soil C accumulation in the biome of 158Tg over 20 years. It has also been assumed that much of this will be accumulated at between 30 and 80cm in depth – a point worth noting as many estimates of SOC sequestration do not consider this much of the soil profile, yet this will presumably be less labile than SOC held in the top 20cm, especially in areas of high erodibility. All this is important, given that the area planted to soybean in Brazil increased by 68 per cent between 1999–2000 and 2004–2005, although credit and profitability problems were expected to slow this trend (USDA, 2005). But that is mainly relevant to the areas of the Atlantic Forest where soybean is significant – the three southern states (Paraná, Rio Grande do Sul and Santa Catarina).

Sugar cane also needs comment here. It has already been stated that above-ground biomass is missing from these estimates, and with sugar cane it could be considerable. The bioethanol question should also probably be considered along with SOC for a holistic view. Brazil pioneered large-scale use of bioethanol derived from sugar cane in the wake of the oil-price hikes of the 1970s, and the first popular car to run on it – a locally-produced Fiat – was introduced in 1979. However, no increase in SOC is anticipated from sugar cane unless 'sustainable practices' are used. Historically, preharvest burning has been used in Brazil to reduce the labour requirement. If preharvest burning is abandoned, a net increase in SOC can be anticipated. Mello et al (2006, p363) quote estimates that abandoning burning of existing sugar cane could in itself accumulate SOC in the biome at a rate of 1.67Tg a year in the top 20cm. Indeed Mello et al (2006, p366) think that total SOC sequestration potential from all practices in the biome over a 20-year period could be as high as 3Pg C, a much higher estimate than Boddey et al (2006) and enough to make a substantial difference to climate change.

But, as Mello et al themselves point out in the case of sugar cane, things are not so simple. New laws are being introduced in Brazil that do ban the burning of sugar cane, but the extra labour required is so great that producers will need mechanized harvesting, and slopes of more than 15 per cent are unsuitable for this. In São Paulo State alone, only about 44 per cent of the land under sugar cane meets this condition. So although the turn to bioethanol will bring some degraded pasture under sugar cane production, elsewhere the trend will be reversed, and it is hard to forecast what the total balance will be. Certainly, from observation in the survey locations, most of the land did not seem suitable for mechanized harvesting (and in one location, in the Zona da Mata in Minais Gerais, sugar cane had already been abandoned for other reasons). In any case, although about a third of the farmers surveyed did have some sugar cane, only a very small number had more than a hectare or so, and no one had more than 4.5ha. Expanding sugar cane was not discussed with the farmers as part of the survey, but in any case it seems unlikely anyone would bother with mechanized harvesting for such small amounts.

The next option in Boddey et al's table is planted forest. They expect a more than doubling in eucalyptus plantations – much distrusted by ecologists, who would prefer to see regeneration to natural forest; but this is not always an alternative. Fast-growing eucalypts can provide useful wood for fuel and fence posts, and were observed in several locations in the Zona da Mata area growing on what appeared to be degraded pasture, but may have been abandoned sugar cane area.

Eucalypts are not new in Brazil. One planter counselled large-scale planting in Vassouras after the coffee disaster (Stein, 1957, p222). Dean (1997, pp232–237) describes how, in 1904, the Paulista Railroad decided to mitigate its consumption of wood for fuel and cross ties by planting forest reserves. This project seems to have been a job-creation scheme for the uninspired godson of a Board member. But he showed unexpected enthusiasm, testing nearly a hundred native and exotic species and plumping within two years for eucalyptus. Their production has not always had happy results, especially in the 1970s, according to Dean: 'Even though prospective planters were advised … to avoid lands suitable for agriculture and to plant on eroded hillsides already denuded of tree cover, they were well aware that eucalyptus grows better on flatlands with good soil not yet subject to erosion – on land, that is, that might, if left untended and free of cattle, grow back to forest,' he says (Dean, 1997, p315).

Boddey et al (2006, pp338–339) argue that there will be little change to SOC stocks from this. But they do not discuss the potential for offsetting cutting of native forest through fuelwood provision and for fencing stakes. At present, using eucalyptus for fencing is a luxury, as Evans (2004, p34) found in São Paulo State: '[M]ost farmers cut Jacaré (*Piptadenia gonocantha*) for fencing stakes without a licence; only the richest farmers can afford to do things by the book, if they wish, by using treated eucalyptus stakes,' he says. But it is hard to know how much cutting of native forest could be offset in this way, and what little was saved would be significant for the preservation of biodiversity corridors rather than carbon. It may also be that sustainable sugar cane might be a better use of valley bottoms than eucalyptus.

The final two options in Table 6.2 are less relevant for this chapter. Coffee is still important in Minas Gerais but has already been much discussed, while cacao is mainly grown further north; in any case, neither is that significant in Boddey et al's calculations. The objective of this section has not been to review all these options in detail (besides, a different range has been used for the field survey). Rather, the objective has been to, first, describe how significant the Atlantic Forest might be in terms of carbon, and, second, to demonstrate the high levels of uncertainty that surround such estimates.

Finally, Boddey et al have drawn up their estimates in the context of agriculture, and therefore do not consider sequestering carbon by giving the surviving native forest a chance to encroach on degraded pasture. This is already permissible under the Kyoto Protocol, unlike pasture improvements or similar, which are not. Moreover reforestation of tropical agricultural or pasture land has huge carbon potential. Silver et al (2000, p394) suggest it could produce an increase in biomass of 6.2Mg/ha a year for the first 20 years and 2.9Mg/ha for the first 80. Tropical reforestation, they say, 'has the potential to serve as a carbon offset mechanism both above- and below-ground for at least 40 to 80 years, and possibly much longer'. Neither is this necessarily from fast-growing species. 'Tropical secondary forests have been reported to accumulate up to 5Mg C/ha/yr during the first 10–15 years of regrowth … and on average have been estimated to sequester 2–3.5Mg C/ha/yr,' they state (Silver et al, pp395–396).

True, the soil carbon figures are not so high, neither are they consistent between former land uses; the authors suggest 1.17Mg/ha/year for unmanaged soils, 0.49 for pasture and just 0.25 for what they term agricultural sites (presumably cropland). Is it better to sequester carbon as SOC in pastures, where it has a chance of being maintained by good management, rather than in secondary forest regrowth that might be cut? Moreover rates of regeneration face unpredictable constraints such as seed rain; although seeds can spread over long distances, they will not necessarily do so. Where seeds are spread by forest animals, they may not end up more than about five metres from the forest edge. Wind-dispersed seeds, too, are less likely to end up in pasture than they are in forest, albeit by a lesser margin (Holl, 1999, pp229, 236–239). Forest regrowth is an unpredictable alternative to pasture improvement and, anyway, it is not an alternative where the land is in use.

There is also no reason to assume that farmers would be amenable to reforestation, even if they were paid to do it. Silvano et al (2005) interviewed nine farmers along the Macabuzinho River and its tributary streams, not far south of the Noroeste Fluminense survey area in Rio de Janeiro State. The farms were mainly pasture. The farmers generally did not like the idea of reforestation even with payment. The authors suggest that cultural factors may have played a part, in that farmers had a tendency to rely on current practices and to view forest as 'useless, less valuable or dirty land' (Silvano et al, 2005, p380). They also speculated that strong official sanctions for cutting native forest made farmers feel that allowing any forest to grow would irreversibly lose them their land.

But what if the occupants have less and less use for the land, their returns are declining, and their aspirations are turning elsewhere? There was no evidence of this among the farmers in the Zona da Mata, but the countryside around the other main location, Itaperuna, had a poor rural infrastructure; the communities in two locations where farmers were interviewed, Cubatão and Serrinha, had trouble maintaining working phone links, and there was no mobile phone coverage. Several farmers commented that no one seemed to care about the countryside. Might farmers one day simply abandon their land, and let the trees grow? It seems highly improbable.

But there is a precedent. The author Bill Bryson, describing the region in which he lived in New England in the 1990s, says:

> *New Hampshire is as big as Wales and is 85 per cent forest. There's a lot of forest out there to get lost in. Every year at least one or two people on foot go missing, sometimes never to be seen again.*
>
> *Yet here's a remarkable thing. Until only about a century ago, and less than that in some areas, most of these woods didn't exist. Nearly the whole of rural New England ... was open, meadowy farmland.* (Bryson, 1998, pp95–96)

The inhabitants simply left, drawn often to land further west that was easier to farm. The extraordinary exodus of the farming population from the eastern United States is well documented. Flinn et al (2005) look in detail at one particular area, Tompkins County in the Finger Lakes region of New York State. Inhabited by the Iroquois in the 18th century, it was, according to land records, 99.7 per cent forested in 1790 (Flinn et al, 2005, p441). European settlement began shortly thereafter and land clearance reached its greatest extent by 1900; there appears to have been little forested area left at that time. But by 1995 it had recovered to 54 per cent.

According to Hart (1968, p435), land abandonment mostly seems to have taken place because of poor infrastructure, the best land being in small patches, lack of paved highways, electricity, or satisfactory schools or churches; while credit and marketing will have been unsatisfactory. All this sounds familiar after meeting the pasture farmers of the Atlantic Forest region. Would these farmers really abandon the land? Possibly not, but as Hart points out, cultural factors should have made it unlikely in the United States also (Hart, 1968, p417).

This book has not seriously considered reforestation because it is concerned with ways in which carbon sequestration can serve farmers, not displace them. But it could be that, like their predecessors in New England, the farmers of the Atlantic Forest will one day just up sticks and move west, to begin the cycle of forest, coffee, and cattle all over again. In that case, it seems most likely the pasture would be replaced with eucalyptus plantations. But perhaps not; the forest may come back after being farmed, as it did before 1500, and once again be the domain of the maned three-toed sloth, black-headed berryeater, solitary tinamou, plumbeous antvireo and the buffy tufted-ear marmoset.

Case-Study Locations

The case study was conducted in 2005 with 37 farmers spread across three main locations. The two most important were in the countryside around the city of Itaperuna in the north-west of Rio de Janeiro State, about five hours' drive from the city of Rio itself; and about 200km to the north in the Zona da Mata region of Minas Gerais State. Both locations are part of the watershed known as the Bacia do Paraíba do Sul; this covers a large part of Rio de Janeiro, and much smaller parts of Minas Gerais and

São Paulo States. Since 2000 the watershed has been under the broad supervision of the Comitê para a Integracão da Bacia Hidrografíca do Rio Paraíba do Sul (CEIVAP).

The north-west of Rio State is known as Noroeste Fluminense (Fluminense being the colloquial name for the state, as opposed to city, of Rio; the two were not unified until 1985). The area was identified for the survey because of the presence of a major rural development project which was GEF-assisted, and had conservation and accumulation of below- and above-ground carbon as one of its objectives.[6] Because of the heavy bias towards extensive pasture in the survey area, it was decided to include farmers with more diverse farming systems within roughly the same agroecological zone; hence the site in the Zona da Mata, although here too there was to a lesser extent emphasis on pasture – a reflection of the Atlantic Forest biome's history, but also of other factors, as will be discussed. Coincidentally, the two survey sites were in or close to the two most deforested sub-basins in the Paraíba watershed: the Rio Pomba (in the Zona da Mata) and the Rio Muriaé (in the Noroeste Fluminense) (May and Geluda, 2004, p20).

Finally, it was decided to include a few peri-urban market gardeners much closer to the city of Rio, near Seropédica in the municipality of Itaguí. This allowed comparison with farmers working on a much smaller scale. This was a small sample and will not be described or discussed in depth, but did give an insight into an entirely different pattern of land-use.

Noroeste Fluminense

The Noroeste Fluminense is an area of 13 municipalities covering 5386.6km^2 in the north-west corner of Rio de Janeiro State, bordered to the south-east by the northern Fluminense region, to the south-west by mountainous areas of Rio, to the north-east by the State of Espírito Santo and to the north by Minas Gerais.

Much of Rio de Janeiro consists of low-lying coastal plain, but inland the landscape is dominated by a mountain range, the Serra do Mar, some of which is forested. The Noroeste Fluminense starts in the mainly sugar cane-growing coastal plain but rises sharply into the interior, the highest point being 1750m (5740ft) above sea level (GEF, 2003a, p4). However, most of it is at nothing like that height. It is also quite dry. Although there is a very small high-rainfall area, annual rainfall in general varies from

2000mm down to 750mm, and it is a water-deficit region with droughts lasting up to eight months (GEF, 2003a, pp91–93).

This has been worsened by the loss of some *matas ciliares* (May and Geluda, 2004, p14); this is a Brazilian term referring to forested areas close to rivers, sometimes translated as gallery forest. Despite that, with one or two exceptions farmers surveyed did not mention water shortages as a constraint. This may reflect the fact that agriculture had already adapted itself to them, the deforestation having taken place originally for coffee. However, deforestation continues even more quickly than elsewhere in the Atlantic Forest biome. GEF (2003a, pp5–6) reports that the State of Rio de Janeiro had a deforestation rate of 16.7 per cent in 1990–2000, far above any other. Despite this, the Noroeste Fluminense, along with the neighbouring northern region, has the largest stands of forest remaining in the state (GEF, 2003b, pp2–3).

Dairy cattle are the main activity, and on-farm land use is largely pasture – 77.3 per cent of farm area, according to GEF (2003b, pp122–126). As elsewhere in the region, this reflects the collapse of coffee, grown widely in the area until the 1950s; GEF (2003b, p128) states that this results from large landowners legitimizing their claim on the land with cattle after that collapse. However, as described earlier, the invasion of exhausted coffee plantations by cattle often reflected the impracticability of anything else.

These are not, today, good pastures. May and Geluda (2004, p26) report that 52.5 per cent of it is what farmers call '*pastagens naturais*', or natural pasture; in fact there will have been coffee there before and the description 'natural' is likely just to mean that they are covered in invasive *colonião*, not more productive grasses such as *Brachiaria*, and have not been planted by the farmer. The neighbouring Norte Fluminense has a higher percentage of planted pasture, and there may be considerable carbon potential in such investment in the Noroeste Fluminense. It was indeed observed during the survey that the pasture covering the low hills of the area was usually unimproved, the difference being fairly obvious; those few that have been planted to *Brachiaria* are a bright green and stand out in an otherwise grey-green, scrubby landscape.

The extensive livestock farming of the area is not highly developed technically. There is limited access to technical assistance (May and Geluda, 2004, p26). In the survey, there were complaints about this from some farmers, whose access to extension services was highly variable.

This is discussed later. Meanwhile, farmers are facing falling prices and competition from other regions, and make few improvements in pasture and feed (May and Geluda, 2004, pp122–126). This is not unusual in Brazil. Maletta (2000, p8) says there is much room for intensification in cattle-raising in place of the current low stocking rates and milk production. One reason may be that 75 per cent of Brazil's cattle are traditional zebu breeds. Most of these are *Nelore* cattle, a breed of Indian origin that takes its name from the city of Nellore in Andhra Pradesh. It was introduced to Brazil at the end of the 18th century. The remainder are cross-breeds (Matthey et al, 2004, p14).

GEF (2003a, pp5–6) claims that 80 per cent of the land in the Noroeste Fluminense suffers from 'moderate to severe' erosion. The low hills do indeed seem subject to frequent gullies, presumably the result of rapid, concentrated runoff. The latter, assisted by removal of gallery forests, has also caused contamination and eutrophication of watercourses with agrotoxins (GEF, 2003a, pp91–93). The high level of deforestation along the banks of the Rio Muriaé contributed to floods in 1997 that affected much of the Muriaé sub-basin, including the main regional city of Itaperuna, leaving 8500 people homeless (May and Geluda, 2004, p20). '[A]reas are seen with soil erosion of catastrophic proportions. Livestock-raising and agriculture are pursued in ways that degrade, and have … to cause a loss of ecosystem services which could be provided with different use of the soil,' say May and Geluda (2004, p30).

The GEF-supported Rio Rural project was designed along PES lines. Typically, a farmer might be paid to use organic manure, or reforest, or practise contour planting on his/her property, up to a defined limit – say a hectare or two per individual. Although there is some large-scale commercial activity, farmers are often not wealthy in Rio, often being family farmers according to the definition of PRONAF – the *Programa de Fortalecimento da Agricultura Familiar*, the programme set up by the Brazilian government to encourage and support family agriculture. The incidence of poverty in rural households in Rio is about 27 per cent, and in Noroeste and Norte Fluminense it is 35–39 per cent in some municipalities; this is about two and a half times the urban level in the state (GEF 2003a, pp5–6). Returns to agriculture are not high, with 90 per cent of rural households having a total income of less than five times the minimum salary, and 80 per cent having less than twice the minimum salary (May and Geluda, 2004, p25).

Leveraging funds for farmers through carbon sequestration is on paper a good form of poverty alleviation. Part of the objective of the survey used in the case study was to see if such an approach would work with family agriculture. It was this type of farmer who was approached as part of the survey. Moreover the technical options used in the questionnaire for the case study were partially those proposed for funding under the GEF Rio Rural project. These subjects are discussed later.

The farmers interviewed were divided between three locations on three sides of the city of Itaperuna. The first, numerically dominant, group were relatively large-scale dairy farmers with farm sizes of about 20–95ha, and were grouped quite close together in the valley of the River Muriaé, about 30km south of Itaperuna. The second group, at Cubatão to the east of the city, were smaller-scale but included one substantial market gardener. Finally there were a number of interviewees at Serrinha, west of the city, with mainly pasture farms varying between about 7 and 80ha.

Minas Gerais: The Zona da Mata

The second location for the case study was some way across the state border in the Zona da Mata (literally, 'forest zone') of Minas Gerais. Minas Gerais borders Rio de Janeiro. The size of France, it spreads across more than one biome but includes a large area of the former Atlantic Forest. Its name is Portuguese for 'General Mines', and in the early years of colonization, it appears to have had little other function. Its initial exploitation was therefore even harsher than that of other parts of the Atlantic Forest. In the 18th century, gold was mined there using methods so damaging to the soil as to make even the roughest pasture, or the most careless coffee production, seem benign. Dean records that it was common to first burn the forest, then divert streams to carry away the topsoil. But some mining was done *a seco* – that is, dry; slaves would carry away 50,000–100,000 baskets of earth to obtain one of gold. According to Dean, the result was to replace forest with 'pockmarked moors'. Inevitably, there was serious soil erosion: 'The denudation of the hillsides caused sheet erosion, giant gullies – called *voçorocas* – silting of streambeds, and flooding that can still be found in the region but are now so generalized and ancient as to appear to be natural features of the landscape' (Dean, 1997, p97).

It was to be some time, however, before this damage was compounded by large-scale agriculture. The Zona da Mata, in the south-east of Minas

Gerais, is contiguous with Rio de Janeiro; its main city, Juiz de Fora, is just 215km (134 miles) from Rio city itself. But in the 18th century it was necessary to force a path through dense forest so that the output from the mines could be brought by mule train to the port of Paratí on the coast of Rio State. At this time the Portuguese Crown forbade the settling of the Zona da Mata, ostensibly because of hostility between Europeans and Indians, but more probably to prevent contraband and tax evasion in a region that was difficult to police (Matos and Giovanini, 2004, p3).

Coffee would be the agent of change. It arrived in Minas Gerais as early as 1809, but took until about 1840 to reach the area around Río Pomba and Ubá, where the survey area was located. With it came rapid change. The number of inhabitants of the Zona da Mata, negligible in 1800, passed 20,000 in 1828, 250,000 in 1870 and 548,000 in 1890, by which time it constituted 7 per cent of the state's population (Matos and Giovanini, 2004, p3).

If coffee had been an environmental catastrophe across the state border in Vassouras, it was scarcely less so in the Zona da Mata. The red-yellow Latossols (Haplic Ferralsols) of the region are not especially prone to erosion but have their limits, especially in fairly high rainfall (Matos and Giovanini, 2004, p3), and the Zona da Mata, which has a mean annual temperature of about 18°C, receives about 1500mm a year, with two to four dry months (Franco et al, 2002, p753). It did not help that the region is mountainous, yet little attention was paid to planting along contours – or use of fertilizer, although information was available by this time, according to Matos and Giovanini (2004, p19). They go on to record that by the beginning of the 1890s, the resultant degradation had started to lower production. In 1877, six municipalities on the Paraíba and Paraibuna rivers had accounted for 90 per cent of Minas Gerais's coffee production, or about 40,000 tonnes; 25 years later this had been cut by three-quarters.

The decline of coffee in the Zona da Mata was accompanied by a conversion of coffee plantations to pasture, as it was in Rio de Janeiro; between 1898 and 1900, export of livestock to Rio more than tripled, with a related increase in exports of dairy products (Matos and Giovanini, 2004, p20). However, the change to pasture was subtly different from that in Rio de Janeiro. The reclaimed plantations were turned to subsistence crops as well as pasture, and the social composition seems to have differed too, with more small-scale family farmers taking over from the coffee barons. It is not clear why this should have been so, but Matos and Giovanini

(2004, p17) report that the end of slavery in the Zona da Mata, instead of resulting in collapse and depopulation, seems to have brought forth a system of sharecropping. In fact, Font (1990, pp308–309) states that sharecropping had always been more important on the large coffee estates in the region. 'This explains why the Mata region involuted to subsistence practices and economic stagnation after the 1930s, with the passing of the coffee cycle,' he says. It may also have helped that the Mineiro government encouraged diversification in the 1900s, reducing taxes on the export of beans, rice, sugar and tobacco, as well as livestock and dairy products.

The keeper of the museum in Rio Pomba told the author that the small farmers had taken over the large *fazendas* when they collapsed, actually

Table 6.3 *Key characteristics of fieldwork locations*

	Noroeste Fluminense	***Zona da Mata***	***Seropédica***
Predominant type	Mainly pasture, but with small areas of crops for household use.	Mixed pasture and vegetable, with a few specialist farms.	Market gardening.
No. of farmers in sample	18	14	5
Mean farm size (ha)*	33.2	19.1	3.2
Average on-farm labour	2.3	3.7	6.6**
Main crops	Dairy cattle, occasionally beef. For household use, cassava, maize and a wide range of fruits and vegetables. Three locations: Muriaé valley (larger farms, almost entirely pasture), Serrinha (similar, but farms typically a little smaller) and Cubatão (smaller, with some market gardening as well as pasture).	Dairy cattle, occasionally beef. Maize and tomato. Wide range of fruits and vegetables, mainly for household use (apart from tomato) but some commercial specialist production.	Mainly rocket, basil, cabbage, spring greens, mustard and various herbs.

* Standard deviation indicates a low confidence level in the mean, but comparison with IBGE figures for the region as a whole indicates that the spread and mean for farm sizes were, in fact, broadly typical. This is discussed further in the next chapter.

** In the case of Seropédica, the mean is distorted by one farmer who employed a lot of labour. Typically, about three family members worked on these plots.

not in the 1900s but in the coffee crisis at the end of the 1920s. Even now, they have more on-farm family labour and a more diversified production system than their counterparts in the Noroeste Fluminense, and are a bit less dependent on pasture. The point is important, because, in the survey, labour was seen in Noroeste Fluminense as a major constraint to carbon-friendly practices, but was less significant in the Zona da Mata. Again, the carbon potential of a region may be path-dependent.

Unlike those in the Noroeste Fluminense sample, the interviewees in the Zona da Mata were all from the same area. This straddled route BR120 between Rio Pomba, which was used as a base, and Ubá. The holdings varied between just 1ha and about 70ha. Although the area has not grown coffee on any scale since the 1920s, it did produce sugar cane until as recently as the mid-1990s. Despite its importance in Brazil, sugar cane is now not much grown in either of the main survey areas. Although some farmers have small quantities of sugar cane for feed, and the Norte Fluminense around Campos dos Goitacazes still has large-scale *fazendas* devoted to the crop, for the most part it has moved south to regions where it can be grown on a larger scale. In the case of the Zona da Mata farmers, a local processing plant had collapsed about 10 years earlier; they ascribed this to mismanagement rather than economics. This had forced them to diversify, and although many had large areas of pasture, others were rotating maize with tomato; another was concentrating on guava. The farmer with just 1ha had turned to production of flower-seeds. One farmer – not interviewed – had started to farm ostrich.

The market gardeners

The small sample from Seropédica were all market gardeners concentrated in a small area about 3km from the town; the latter is only an hour or so's drive from the centre of Rio de Janeiro but has a distinct identity of its own and is the site of the rural campus of the Federal University of Rio de Janeiro (UFRJ). They had obtained their plots from the Instituto Nacional de Colonização e Reforma Agrária (INCRA) some years earlier; they had paid for them, but, they said, not much. They were typically 2–4ha, worked with family labour and were of limited means; but one, an immigrant who had arrived from Portugal in 1958, had 22 employees on 6ha and was clearly running a large commercial operation.

The Next Step

The two main areas reviewed are of great interest for the carbon sequestration potential in their degraded pastures. But there are also other reasons why they are a good test case for carbon. One is that the area looks capable of supporting a more diversified and sustainable agriculture than it does now. Another is the short distance (by Brazilian standards) of the sites from large and medium urban centres such as Rio de Janeiro and Juiz de Fora; as discussed in the previous chapter, this may be relevant to REDD+. The Seropédica sample is small, and less significant in carbon terms, but its inclusion does make the group more typical of the region in general.

However, none of this means that the farmers would want to sequester carbon, that it would fit into their farming system, or that the practices demanded would make sense given the broader farming economy. If these conditions are not met, sequestration activities may not be adopted, or if they are, the sinks created may not long outlive the intervention. There is ample evidence of this in the literature.

Moreover, before deciding that carbon can be recovered in a biome, it was as well to understand how it got away in the first place. This is important for two reasons. First of all, the progression from forest to coffee to cattle defined the agroecological conditions seen there today; an understanding of that process clarifies what might need to be done to sequester and retain carbon in the region. Second, if the fundamental drivers of resource use have not changed, should a CDM-type project be trying to change them? The survey may help to answer this.

Notes

1 There are many definitions of a biome. In their discussion of Andean forest ecosystems in Colombia, Armenteras et al (2003, p247) cite that given by Hernández (1990): an assembly of ecosystems with similar structural and functional characteristics. This is broadly applicable here. The choice of the word should not be taken to imply any particular ecological approach in this chapter or the one that follows.

2 Perhaps not quite the first time! The explorer Robert Marx claims to have recovered amphorae from a 2nd-century Roman wreck in Guanabara Bay in 1981.

3 Degraded pasture is here defined according to Boddey et al (2004, p389), that is, that they 'support very low stocking rates, show low plant cover, are invaded by non-palatable native species and often densely populated with termite mounds'. From observation, many also show signs of erosion, either in small localized patches or gullies. Sheet erosion will also be present, but is harder to detect. Because of this, and because localized erosion may or may not affect productivity, the omission of erosion from Boddey et al's definition is reasonable, but where pasture is degraded, erosion and its attendant loss of SOC will usually be present.

4 The title of Warren Dean's book, *With Broadax and Firebrand*, is derived from one of Vandelli's Brazilian students, another Mineiro, named José Vieira Couto, who wrote in 1799 that the Brazilian farmer had 'a broadaxe in one hand and a firebrand in the other' (Dean, 1997, p138).

5 A *braça* was equivalent to 2.2m; 3000 were equal to a *légua*. Some caution is needed with older Brazilian measures, which may vary from state to state; some (for example *alquieres*) still mean slightly different areas to farmers in Minas Gerais, São Paulo and Rio de Janeiro.

6 The description of Noreoeste Fluminense in this chapter draws heavily on the project documentation prepared for GEF and for the Brazilian project management.

7

The Sceptical Farmer

Why do farmers resist taking advantage of knowledge that would improve their production capacity, their yields and the quality of their lives? Their clinging to traditions is the reason most often expressed by technical staff and the local elite ... (GEF, 2003a)

As stated in the previous chapter, the objective of the survey was to uncover the constraints to carbon-friendly practices and see what their practical potential might be. The difference between this and the theoretical potential is huge. Bridging some of this gap may depend in part on absorbing the messages of previous agricultural development projects that had nothing to do with carbon.

It should by now be axiomatic that what looks sensible on paper does not always suit the farmer. This applies to the study area as much as anywhere, as the GEF project brief for the Rio Rural project in the area confirms. 'The issue of resistance to the introduction of new technologies and practices for crops and agricultural management is a widely-shared perception among those who provide technical assistance and rural extension services to farmers and residents in rural areas in the North and Northwest Fluminense,' it states (GEF, 2003a, p134). It goes on to cite the planting of crops on land cleared by burning, the use of traditional low-yield crops, extensive cattle-raising and monocropping.

A similar view was expressed by the technician employed by one of the farmers' associations in the Muriaé valley, who told the author that he used to suggest that people try coffee or fish-farming, or pigs; one might use the waste from one to feed the other. This is established elsewhere in Rio de Janeiro State. Raising rabbits is also promising because the meat is sought after. But, he said, farmers are resistant to such proposals because the farmer feels comfortable with what s/he has always done. Such resistance has historically infuriated those who claim to have the farmers' best interests at heart. This has especially been the case with soil erosion, which arouses emotional responses in some.

Carbon: The Challenge of Non-adoption

In fact, farmers may reject measures because they are uneconomic or disruptive. As Pretty and Shah (1999, p9) put it: '[M]ost soil and water conservation programmes have begun with the notion that there are technologies that work, and it is just a matter of inducing or persuading farmers to adopt them. Yet few farmers are able to adopt whole packages of external technologies without considerable adjustments in their own practices and livelihood systems.'

Even if they can, they may not want to, especially so when the notional gains from such technologies are not as great as they appear. Tropical soils can often tolerate considerable soil loss before it is worth addressing (Hellin and Haigh, 2002, p240). Even where it cannot sustain such losses, the quality of the soil may not justify its conservation. Blaikie (1989, p27) uses the example of a gully appearing in a farmer's field in Lesotho: 'It is more cost-effective … to find employment in the South African mines than to expend … scarce resources patching up land of such low productivity.'

Where measures are worthwhile, they will still not be taken if other measures promise more. Tiffen (1996) points out: 'Within the farm, capital first tends to be invested first in the most profitable part of the enterprise (e.g. in a terrace for growing tomatoes, or a plough to expand cultivation) and to spread gradually to investments giving a lower return

Table 7.1 *Farmers' reasons for non-adoption of soil-conservation technologies*

- Technologies demand changes that are alien to the farming system
- No secure access to land
- Labour costs
- Technology may not stop soil loss, or increase yields
- Physical earthworks may not in themselves increase productivity
- Technologies force land out of production
- Farmers aren't convinced that erosion is a key problem
- Resistance to 'top-down' programmes
- Technologies exacerbate other problems – waterlogging, weeds, pests, diseases
- Farmers don't feel they own the technologies
- Increases in risk or debt

Source: adapted from Hellin and Haigh (2002, p236)

(e.g. the improvement of grazing land).' Even where soil conservation is the first priority, environmental concerns will not dictate the choice of soil to be conserved. Lu and Stocking (2000a, pp150–151; 2000b, pp163–164) report that farmers on the southern Loess Plateau in China find it more economic to limit their investments in soil conservation to the better soils; as in any industry, it is better to invest in a productive asset than an exhausted one. As Lu and Stocking say (2000a, p150), the return of residues to the soils might be part of good management, building up organic matter, but is an expensive process if erosion processes are not stopped. This is logical; one would not ask a manufacturer to refurbish obsolete or unprofitable machinery instead of updating newer plant that works better.

All of this is well known to agronomists who are either from the developing world, or have worked there. However, decision-makers in the broader environmental arena would not necessarily have been exposed to this type of literature. It may be that they will see what appears to be a win-win solution – soil conservation for higher productivity *and* climate-change mitigation – and will not understand why farmers' cooperation cannot be taken for granted. In any case, incentives have a doubtful history in soil conservation. Could overriding farmers' judgement in this way be counterproductive?

This is an important point, as carbon sequestration initiatives would effectively be incentive schemes, and there is a large literature on them (for Ethiopia alone, particularly regarding the food-for-work (FFW) programmes of the 1980s, see Herweg, 1993; Pretty and Shah, 1997; Pretty and Shah, 1999; Bewket and Sterk, 2002). There is now plenty of evidence that incentives, or at least direct inducements such as cash payments or FFW, in soil conservation often do not work (L. Lewis, 1992; Herweg, 1993; Kerr and Sanghi, 1993; Sombatpanit et al, 1993; Stocking, 1993).

This should not be oversimplified. As Sanders and Cahill (1999, p14) explain, incentives can be indirect as well as direct. Indirect types include fiscal and legislative measures, guaranteed input prices, suitable land tenure, equipment, marketing, storage, education and more. They can include extension services; these can be crucial where farmers can see that something needs to be done but are not sure what or how, and find it hard to get advice, as in the case of some of the Brazilian farmers described in the case study. Even direct incentives need not be direct payments or FFW; they might be in kind, as agricultural implements, livestock, trees or seeds (Sanders and Cahill, 1999, p14). Enters (1999, p35) adds others, including tax credits. 'For example, pollution taxes and tax credits for investing in soil conservation benefit land users directly and enable them to shift production towards more sustainable practices,' he states. This could include the provision of subsidized credit.

Some caution is needed here, however; Pagiola (1999a, p23) warns that this can be a perverse incentive leading to measures that the farmer did not really want, and will abandon later. He recounts how one project in El Salvador extended subsidized credit so that farmers could take conservation measures: 'In practice … farmers appear to have been undertaking conservation measures so as to gain access to credit, making conservation a *cost* of obtaining credit rather than a benefit of doing so,' he says. A related point is made by Enters (1999, p37), who points out that farmers may regard incentives simply as payment to do something they would not do otherwise, just as most of us do when we go to work in the morning. So he finds it quite natural that, when incentives cease, the work paid for will too. 'Since they view it as a wage payment it should not be surprising that once the payments stop farmers drop their tools. After all, aren't most of us paid for the work we do?' he says.

Subsidizing inputs can also have unpredictable results. Pagiola (1999b, p49) suggests that subsidizing fertilizer – an option that is significant in the case study, as will be seen in the next chapter – could 'reduce … nutrient depletion in one area, but discourage farmers from adopting terraces in another because artificially inexpensive fertilizer provides a more profitable means of maintaining production'. De Graaf (1999, p113) suggests that subsidies might discourage farmers from using locally available organic alternatives and also reports that they have encouraged overuse of fertilizers in Indonesia.

Of course, that does not mean that, having been induced to take conservation measures, farmers will not find them worthwhile and decide to keep them. Giger et al (1999, p270) point out that an incentive scheme can help farmers bridge the gap between initial investment and return. It may even be helpful when no such gap exists, simply because farmers are so risk-averse. Giger et al report the results of a survey by the World Overview of Conservation Approaches and Technologies (WOCAT) project in 15 countries of Southern and East Africa. Of the projects reviewed, 80 per cent used some kind of incentive, and they reported mainly positive outcomes. True, a significant minority of respondents did see incentives as moderately or very negative in the long term, making it hard to persuade farmers to act without them, or creating dependency, or distorting relationships with extension agents. However, about 70 per cent thought the long-term negative impact of incentives was low or non-existent (Giger et al, 1999, pp255–256). It does seem that direct incentives, carefully designed, can have positive outcomes.

Kerr et al (1999, p306), who analysed soil and water conservation (SWC) investments between 1987 and 1997 in 29 villages in Andhra Pradesh, make a number of recommendations for ensuring this. They include:

- Farmers must pay a significant proportion of the cost of any technology (to ensure that they do want it). 'Only if they have to make a real contribution will there be any guarantee that they will … [maintain] technologies that are introduced,' they say.
- Access to credit is needed for improving plots, but without tying them to any particular technology – farmers might reject it.
- A sliding scale should apply, with lower subsidies for improving the better land, because farmers might improve this anyway. (This has relevance to the additionality problem discussed in Chapter 5).

However, their first and perhaps most important recommendation is that investment must be in projects farmers want. 'If they do not want vetiver grass, for example, there is no point in forcing it upon them as they will refuse to maintain it,' they say (Kerr et al, 1999, p305).

This implies that farmers should be involved in the selection of technologies. Bewket and Sterk (2002, pp197–198) suggest that in the Ethiopian case they reviewed, which concerned inappropriate soil conservation measures in East Gojjam, the farmers needed to be involved in the design of the measures; this would have prevented inappropriate solutions. This implies programmes that are an interactive process between the farmers and the funding agency (in this case, the government; but it could also have been the World Food Programme or a similar body). Although the authors do not say so explicitly, this would have clashed with the administrative ethos underlying the project, as revealed by the development agents who had pressured the farmers into participating; they apparently believed that the work 'was not for the sake of conserving the farmers' soils and lands, but to meet demands of the government's five-year development programme' (Bewket and Sterk, 2002, pp193–194).

So the project was target-oriented. This raises a worrying prospect: government or regional agriculture ministries promised a certain cash income on the basis of so many tons of carbon sequestered, or simply conserved – at any rate, maintained above a given baseline, perhaps through soil-conservation programmes all too similar to the ones described above. Not only would these not necessarily maintain carbon above the baseline; they might actually lose it. Pretty and Shah (1999) comment that: 'Local people whose land is being rehabilitated have found themselves participating for no other reason than to receive food or cash. Seldom are the structures maintained, and so conservation works deteriorate rapidly, accelerating erosion instead of reducing it' (Pretty and Shah, 1999, p10).

There is, therefore, clearly a need for a methodology that will allow farmers to review the management practices proposed for carbon sequestration at the planning stage. This must be done with care; it is not helpful if farmers are simply asked whether they would like to reduce stocking rates, build bunds, plant contour strips or whatever, as a simple 'no' will convey little. The methodology must ask farmers to assess the carbon-friendly practices in a manner that is constructive for the project designer. If they do not wish to adopt a practice, is it because it has no benefits for them, or would require too much labour to maintain? In such cases, there is little point in throwing money at the problem, as the practice will not be maintained when one has ceased to do so. However, if it is a matter of initial cost or technical knowledge, it might be practicable.

A Methodology for Assessing Options

The methodology chosen for this must reflect the role of soil carbon. It would be easy simply to regard it as a commodity, which farmers can choose to produce or not depending on the price asked for it. However, carbon is different because of its global significance. As Brown and Corbera (2003, p8) put it: 'The development of carbon markets may privilege global claims over those of other users and scales.' In other words, the need to mitigate climate change may be seen as such a high global priority that it could sweep aside the needs of local stakeholders such as farmers. Hurni (2000) distinguishes between 'internal' and 'external' stakeholders; internal views are 'those derived from local knowledge and experience, while external views are based more on knowledge of a global nature, derived from scientific research and education.' Broadly speaking, internal stakeholders may be expected to have a *use* value for the resource, while external stakeholders will have an *existence* value.

These viewpoints must be reconciled if any carbon project is to be sustainable in the long term. Farmers do not practise soil and water conservation for its own sake, or because external stakeholders think they should. They do it because it raises their income or nutritional status. This is a point that has emerged from the WOCAT project, mentioned earlier, which has collected and reviewed examples technologies for sustainable land management devised by the farmers themselves. Critchley and Mutunga (2003, p160), discussing 18 such projects in Africa, state that all the innovations reviewed had been mostly developed for improved production or income, even though runoff control or other aims were sometimes pursued as co-benefits. 'Here is a central message: conservation is never divorced from production in the eyes of the innovators,' they say.

In light of this, the approach taken was to ask farmers to weight various 'carbon-friendly' management options against different criteria. However, those criteria did not include carbon sequestration. That is an output desired by external stakeholders, and the farmers' priorities lie elsewhere.

The farmer was asked how s/he rated a management practice, on a scale of 1–10 against certain management objectives s/he might have. Thus the respondent was presented with 11 management practices that might be used to enhance the carbon sink, and asked: how much they liked them; why they were using them (from a multiple choice menu of reasons); if they were *not* using them, how much they would like to; and finally, the extent to which they thought that practice was good for the soil.

The initial rating scheme of management practices is shown as Table 7.2. The rationale for picking individual practices is discussed on a case-by-case basis in the next chapter.

This approach – of producing a 'basket of technologies' which farmers can prioritize, and reject those they do not want outright, rather than a prescriptive work programme – reflects many years' post-colonial experience with transfer of technology in agricultural development. Part of the reason why soil-conservation technologies have been rejected is a failure to appreciate that they are very site-specific and, even more important, *farmer*-specific.

For example, a technology package might include additional fertilizer; in that case, adoption would be constrained by the amount of labour required for its application, so proximity and ease of access to the field

Table 7.2 *Questionnaire for farmers' ratings of carbon-friendly practices*

Option	***How much farmer likes this practice***	***Why is s/he using it? Use letters A–G according to key below***	***If not using, how much s/he would like to***	***Extent to which the practice makes the soil stronger or weaker***
Inorganic fertilizer				
Green manures				
Minimum tillage				
Contour planting				
Vegetative strips				
Sustainable rotations				
Pigeonpea				
Leguminous tree species				
Fruit trees				
Pasture improvement				
Organic agriculture				

A: Feed family; B: Feed animals; C: For sale in the market; D: Maintain soil fertility; E: Prevent soil erosion; F: Provide fuel for household; G: Other would matter, as well as the cost of the input. The same applies to other inputs. Bationo et al (2007, p16) quote Prudencio's (1993) figures for SOC content in home gardens, and near and distant fields in Burkina Faso; the further the field, the lower the SOC content. 'Usually, closer fields are supplied with more organic inputs as compared to distant fields due to the labour factor,' they say. Availability of labour to carry the manure might depend on the number of children in the household. Where use of manure is recommended, the number of animals owned by a household could also determine feasibility (Williams et al, 1995), or could change the balance of profitability. So farmers prefer a basket of technologies from which they can take those they want and try them out. Indeed, some would see this as part of a process by which farmers actually *develop* the technology; as Sumberg and Okali (1997) put it: 'High-quality agricultural extension should be defined in terms of its success in fuelling farmers' own search for site- and situation-specific solutions and opportunities.'

In carbon projects, it is true, there will be pressure for the management practices to be predetermined for accounting purposes, so this model might not apply. However, it will be better if farmers are compensated for carbon gains without specifying the technologies too closely. This would make monitoring and verification more dependent on actual measurement of carbon; as discussed in Chapter 6, it is much easier to use stratified accounting based on the projected accumulation of a practice, and simply ensure that that practice has been adopted. However, farmer-specificity is important. Why would a farmer have the same enthusiasm for (say) contour planting as his neighbour when they have completely different topography? The 'basket of technologies', from which they can choose, is therefore essential, and they must be able to develop and adapt the chosen technologies for their own farm.

A further reason for respecting heterogeneity in this way emerges from the cost–benefit analysis performed by Tschakert (2004, 2007). As stated above, one should have reservations about cost–benefit analysis in PES, as it is predicated on the assumption that farmers can be paid to do some things they would not otherwise do, and the history of soil conservation suggests otherwise. However, Tschakert looked at the profitability of practices without as well as with carbon payments, which implies that the results are an indicator of sustainability. Her work in the Old Peanut Basin of Senegal remains the best field analysis of what a carbon sequestration project in agriculture, as opposed to forestry, would actually do. She found that the costs of adopting a given practice tended to be higher for resource-poor farmers, to the extent that two-thirds were more profitable for better-endowed households and only one held more advantages for poorer ones (Tschakert, 2007, pp78–79). Similar conclusions were reached from a very different agroecological zone when the possible impacts of carbon sequestration were reviewed in a slash-and-burn system in Panama (Tschakert et al, 2007). It is hard to see how this can be completely eliminated, as 'investing' in carbon will often incur opportunity costs from the forgone use of an asset, and as poorer households will have fewer assets, that used will for them constitute a larger percentage of household income. But the impact can be reduced with a 'basket' of technologies that permits diversity of asset use for sink management.

The management options for the survey 'basket' were selected in two main ways. First of all, they included some of the options relevant to carbon sequestration included in the GEF Rio Rural project. These were not listed in any of the formal project documents, but the project management kindly shared them with the author. They included organic fertilizer, contour planting, vegetative strips, pasture rehabilitation and sustainable rotations. The other options were picked following discussion with staff at EMBRAPA Agrobiología, the soil fertility and chemistry arm of the Brazilian state agricultural research organization, EMBRAPA (Empresa Brasileira de Pesquisa Agropecuária). These included the use of *inorganic* fertilizer, which it was felt might be a more realistic option for some farmers. The Rio Rural list also included agroforestry; this was omitted as it was seen as hard to define, but the use of fast-growing leguminous tree species was included instead, as recent experiments by EMBRAPA had suggested very high carbon sequestration potential and potential to fix nitrogen in degraded pasture. The rationale for each practice is discussed further in the analysis in the next chapter.

The table was intended to indicate whether or not the farmer actually *wanted* to adopt the sustainable practices; if they did, but had not done so, it was worth finding out why not. This would suggest the options for carbon funding that met with approval from farmers.

But it would then be necessary to indicate whether practices went unadopted because of a specific bottleneck (lack of seeds, technical knowledge), or whether there were systemic constraints (labour, lack of a market). Whether or not something is a systemic constraint, or one that could be addressed with a little extra money, is obviously dependent on local factors; sometimes a market can be created simply by building facilities, or labour recruited with slightly higher wages, but if the consumers or the labour are simply not there or are too far away, there is a systemic constraint. In the case of the Noroeste Fluminense survey area, for example, there was a widespread perception that there was no longer any obvious nearby source of labour, even if money were made available. A simple bottleneck might be addressed by carbon funding; but a systemic constraint probably could not. However, its existence might reveal something useful about the relationship between carbon management and the broader economy.

To this end, the second major table, shown as Table 7.3, asked farmers to rate constraints to adoption, again on a scale of 1–10. The list of constraints was initially drawn up by including the sorts of constraints that would normally face technology adoption, and an initial pilot survey of some of the Noroeste Fluminense farmers suggested that they were the right choices. One option, 'Displaces more profitable activities', seemed less relevant but was kept in. In fact, this constraint was probably covered by farmers' responses to labour as a constraint, as a limited supply of labour could only be directed at a given number of activities.

The rating of practices and constraints on this scale presents a challenge in that one farmer might express (say) a serious constraint with a value of 6, while another might say 9 although the weight of the constraint was really the same. Tompkins (2003, p12) warns: 'Different voters have different personal value scales so that even though each may allocate, through a voting procedure, the same numerical value to different criteria, they may in fact have different intensities of preference for them.' This was sometimes evident in the survey; for example, one farmer gave ratings of 7 and 8 out of 10 to two constraints, despite having said earlier that they were not really a problem.

This difficulty has never really been resolved (Tompkins, 2003, p12; Brown et al, 2002), but can be mitigated by using appropriate types of data. Thus Brown et al (2002, pp101–104) describe four types of data:

- *Nominal data.* The respondent is asked simply to state whether individual criteria are important to them or not. Nominal data are easy to collect, but convey little information.
- *Ordinal data.* The respondents rank the criteria in order of importance. This is less ambiguous, but still does not reveal how emphatic the preferences were.

Table 7.3 *Questionnaire for farmers' ratings of constraints to adoption of carbon-friendly practices*

Option	*Initial cost*	*Does not generate useful income or products*	*Displaces more profitable activities*	*Lack of technical knowledge*	*Demands too much labour*	*No market for products*	*Other*
Inorganic fertilizer							
Green manures							
Minimum tillage							
Contur planting							
Vegetative strips							
Sustainable rotations							
Pigeonpea							
Leguminous tree species							
Fruit trees							
Pasture improvement							
Organic agriculture							

- *Interval data.* A matrix is drawn up in which a criterion can be marked as very, fairly, or slightly important to a respondent. This forces respondents to consider the relative strength of their preferences, which ordinal data does not. On the other hand, the gaps between 'very', 'fairly' and 'slightly' (the intervals) are fixed, so the expression of this relative strength is crude.
- *Ratio data.* The respondent is asked to allocate 100 points across the criteria. Thus, out of four criteria, an official dealing with the CDM might allocate 40 points to carbon sequestration, 35 to availability of livestock products and 25 to provision of clinics, but none to marketing opportunities for a certain product.

Of these options, ratio data most accurately reflects respondents' views. It also provides more opportunities to 'unpack' the data and to understand who is saying what, and why. But it also limits the number of criteria that respondents can be asked to weight, as it increases cognitive complexity. It would also have caused great difficulty in the farmer interviews, where farmers sometimes had a view on one practice or constraint but not another; they could therefore not have allocated ratio data between those options. The cognitive complexity would also have been a challenge – to the interviewers, as much as the farmers. Interval data would deal with this, but as Brown et al point out, relatively wide fixed intervals would give very coarse data.

So a scale of 1–10 was used. Perhaps inevitably, farmers sometimes either did or did not see a practice as attractive, or a constraint as serious, and allocated either a zero (or 1) or a 10. If this were to occur too often, the data would effectively become ordinal data and would say much less about (say) the extent of a constraint to a practice, so the more differentiated the ratings for a given question, the more informative the data.

The Interviews

Interviews were semi-structured and began with the questionnaire, with discussion afterwards. While the sections on carbon-friendly management practices and on constraints to those practices were the core of the questionnaire, background data were also gathered on the farming system in order to put these in context. Not all of this proved relevant; what was, is presented in Chapter 7. The respondents were asked:

- Number of persons working on-farm. Had these included children, those under 14 would have been entered as half a unit in line with IBGE practice, but the labour was mostly adult.
- Area of crop production for a number of crops, including maize, soya, cotton, sugar cane, cassava and green manure. In practice only maize, rice, sugar cane and cassava were common and were mostly grown on very small areas for household use.
- Area of pasture and forest.
- Number of head of dairy and beef cattle. In some cases farmers gave the figure for dairy plus an overall figure that included heifers.
- Daily egg production.
- Area of fruit production.

The farmers were then asked the core questions on the practices and their constraints. Afterwards, they were asked what were the principal constraints they faced as farmers in general, being asked to rate (out of 10) six options: lack of credit, labour shortages, insufficient land area, pests and diseases, lack of marketing opportunities and soil erosion. Besides rating the constraints, farmers frequently had comments to make about one or more of them and were encouraged to expand on their answers.

They were also asked whether they were members of a farmers' association. This is potentially important as smaller farmers would not necessarily be able to exploit the value of their carbon on an individual basis. In Noroeste Fluminense, all were members of an association, but these were very local and were really neighbourhood groups. In the Zona da Mata, about half were members, mostly of a local community association but two or three of other groups, including trade unions (one was a member of the Sindicato Trabalhadores Piraúba).

It is not clear how representative the sample was in this respect. It was thought unlikely that Brazilian farmers would agree to be 'cold-called' by a researcher they did not know, not least because they would be concerned about the possibility of any data reaching the tax and regulatory authorities. They were therefore approached through individuals in the community, and in the Zona da Mata and at one of the Noroeste Fluminense sites, these were the organizers of the local associations. So there is almost certainly a sample bias towards members of associations. However, although at the other Noroeste Fluminense sites, the introduction was through the Itaperuna extension agent, the farmers were nonetheless members of their community groups. Even in Seropédica, where no farmers were members and the introduction was through a member of a local church, two of the farmers had been members of farmers' associations or unions in the past.

Finally, the farmers were asked how long they had been farming – this ranged from just one year to 65 years. To finish the interview, farmers were encouraged to talk about the state of the local farming economy and whether they thought yields were rising or falling. Inevitably, some were more talkative than others.

As indicated above, the choice of farmers to be interviewed was to some extent dictated by the range of contacts available. However, this was not the only criterion. The Noroeste Fluminense area was originally selected because of the GEF/Rio Rural project in the area, which had carbon sequestration as one of its planned outputs; on-farm carbon was therefore relevant and topical. However, there was also a preponderance of extensive pasture with few other activities on a commercial scale. The INCRA/FAO (2000) survey of family farming in Brazil suggested that such a bias could indeed be expected in the region, but not to that extent. So it was decided to rebalance the survey by including the Zona da Mata farmers, who also used much pasture but were less dependent on it. It was also noted that there were a large number of very small plots in the region, and a group of these were identified near Seropédica. The resulting mix and its relationship to the profile of family agriculture in South-East Brazil is discussed further in Chapter 8.

The nature of the constraints that farmers identify to the management practices needed may help identify whether carbon can be restored to the system through CDM-type, pump-priming funding, or whether they are systemic. Is the use, or abuse, of soil carbon path-determined, locked in by history and the global economy? And if it is, what should be done about it?

8

The Farmer's View

> *We do not want your 'Marxist' NGOs coming to tell us what we should or should not be doing with our rain forest. Please send us some experienced farmers and agriculturalists. There is so much land that has already been cleared and is run very extensively. ... Show us how to farm it more intensively in an environmentally friendly manner and then there will be less pressure on the forest.* (Professor at ESALQ (Escola Superior de Agricultura Luis de Queiroz), Piracicaba, quoted in Roberts, 2006, p24)

The case-study survey is presented in this chapter. It combines the results with analysis of the carbon-oriented management practices in the context of the Atlantic Forest biome. Where practices cannot easily be separated from each other (for example, organic agriculture and green manure), they are considered together. The amount of detail given for each practice varies according to how relevant a practice was, in terms of both its carbon sequestration potential and its likely adoption.

First some basic information about the farmers surveyed is given, followed by a comparison of the sample with the norms for the region – this is clearly important if the results are to be useful on a wider scale. The farmers were also asked what constraints they faced in general, as this can help illuminate why there are certain perceived constraints to the proposed management practices. In the following chapter, the results are discussed and some preliminary conclusions from the case study are presented.

The Sample and Farming System

Farm size for the sample as a whole varied widely; the mean was 23.8ha but the range was 1–94ha, with a standard deviation from the mean of 25.3. (The smallest plot was actually in the Zona da Mata, not the INCRA settlement.) There was a marked predominance of pasture in the samples

– around 80 per cent of farm area in Itaperuna, and just over 50 per cent in the Zona da Mata. In the case of the latter, this is misleading, as some farmers were more interested in vegetable (especially tomato) production, and three had no pasture at all. But in the Itaperuna region, even the few farmers who did devote much of their effort to other activities, still used much of their land as pasture.

Of those who did have pasture (about three-quarters of the sample), there was an average of about 24 dairy cattle and just under five beef cattle. The stocking rate is therefore about 1.2 head per hectare, compared to a national average of 0.9 head per hectare (Carvalho, 2006). However, the national figure is also misleading as a comparison, as beef cattle greatly outnumber dairy cattle in Brazil, whereas in the survey area it was the other way around; moreover much cattle production is on savannah, a very different system. In some parts of the Atlantic Forest region, the stocking rate might be much higher; in the experiment described by Boddey et al (2004) in Bahia and discussed later in this chapter, it was increased from two to four animals per hectare (beef cattle). For the region, the stocking rate in the survey area was probably quite low. (It excludes heifers as some farmers did not regard them as part of the dairy herd and did not mention them.)

Other activities took up much smaller areas. Sugar cane averaged about 0.9ha per farm; in practice, a small number of farmers were growing between 2 and 4.5ha, while the majority grew little or none. The average area planted with maize was just under a hectare, but again this is deceptive. Farmers sampled in the Zona da Mata region of Minas Gerais averaged just under 2ha, while elsewhere a few farmers had maybe a hectare for household use and the remainder none at all. In fact, even in Minas, most of the maize was being grown by four or five farmers who also grew tomatoes; the maize takes advantage of residual fertilizer that is worth using for a high-value crop like tomato, but would not be used on maize alone. Three of these farmers were growing tomatoes on a commercial scale, with areas of 2, 2.5 and 4ha.

Other crops included beans, grown by almost a third of the farmers but mostly only for household use; cassava was similar, being grown on nearly half the farms but almost always on areas of less than a hectare. Cassava-derived dishes remain popular in Brazil – the main road from Rio de Janeiro to Vittoria is lined with restaurants advertising country cooking, with dishes such as *bolo de aipim*, balls made from cassava flour. But the labour requirement for processing cassava is quite high. A small number of farmers had some rice, but only one was growing more than a hectare; she had 4ha. As she was one of the few farmers who was renting, it may be that this was a safer bet than investing heavily in cattle, of which she had few. One farmer had 9ha of tobacco, but almost no one else had any at all.

Almost everybody grew a few bananas and oranges for household use. There were also many other fruits, vegetables and herbs grown on very small areas, again probably for the household; they included lemons, spring onions, coconut, passion fruit, guava (one farmer was producing this commercially), coconut, carambola (*Averrhoa carambola* L., a fruit originating in South and South-east Asia), tangerine, mango, pineapple, blueberry, caja (*Anacardium giganteum*, a fruit native to the Amazon), avocado, lettuce, cabbage, radish and more besides. A summary of the main characteristics of the three main survey areas may be found in Table 6.3 (p154). For the market gardeners in Seropédica, with their scarce space, emphasis was on high-value products that could be grown in a small area. These included rocket, basil, cabbage, spring greens, mustard and various herbs.

A number of farmers had some forest on their land. Although the Portuguese word used in the questionnaire was *mata*, meaning simply forest, some farmers appeared to understand this as *reserva* – protected forest. However, the only other forest they could have had would have been that they had planted themselves, and only two had done this. So the average of just over 1.5ha was presumably Atlantic Forest, although it may have been secondary regrowth. Again, just under half the farmers had no forest at all, so the stands that did remain would have been bigger than the average suggests. One farmer in the Zona da Mata had 10ha and another 17ha. Both were less reliant on dairy cattle than the average; this may indicate a link between lesser reliance on cattle and a greater willingness to accommodate forest. However, this would be hard to prove without establishing who had cut *reserva*, when, and why. As it is illegal, this line of enquiry would have been unethical and might have been misleading.

The Sample in Context and South-East Brazil

If the sample is atypical, then the findings cannot be meaningfully scaled up, so basic indicators for the sample were compared with those for the region. For the latter, it draws heavily on the census data compiled by the Instituto Brasileiro de Geografia e Estatística (IBGE) in the *Censo Agropecuário* of 1995–1996.[1] There will be many agroecological zones even within a state; Minas Gerais for example includes the Cerrado, which is outside the Atlantic Forest biome. So comparisons were made at municipality as well as state level; however, only the more important comparisons are presented here.

It does appear that the sample is broadly typical of the mixture of extensive pasture, mixed family farming and peri-urban agriculture found in South-East Brazil. The first comparison is the status of the farmers on

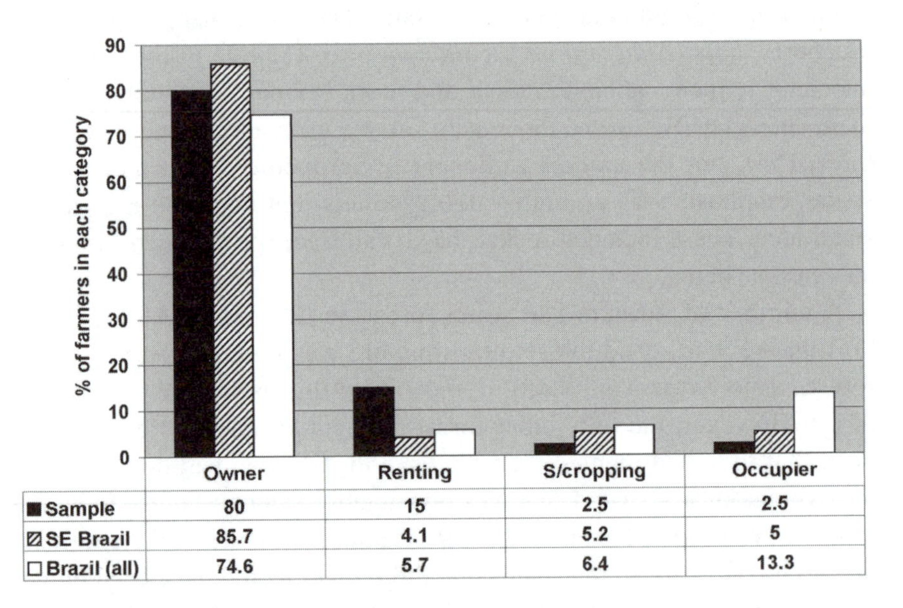

	Owner	Renting	S/cropping	Occupier
■ Sample	80	15	2.5	2.5
▨ SE Brazil	85.7	4.1	5.2	5
□ Brazil (all)	74.6	5.7	6.4	13.3

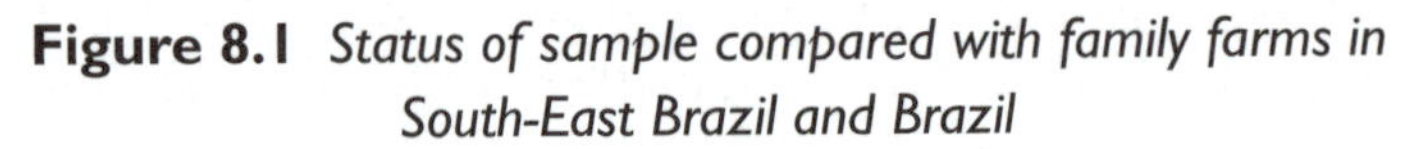

Figure 8.1 *Status of sample compared with family farms in South-East Brazil and Brazil*

Source: sample data; INCRA/FAO (2000, p22)

Note: The sample data here includes one employed farm manager/labourer under 'Occupier'. The INCRA/FAO data was adapted from the IBGE *Censo Agropecuário* of 1995–1996 and the assumptions on different status types will therefore be the same. However, INCRA/FAO are looking only at family farming (*agricultores familiares*). As the sample did not include any farms that fell outside this category, the comparison is acceptable, but in the other graphs in this chapter the data has usually been drawn direct from the IBGE survey and includes all farms. The area defined as 'South-East Brazil' covers the States of Espíritu Santo, Minas Gerais, Rio de Janeiro and São Paulo.

the basis used by the 2000 study of family farmers compiled jointly by INCRA and FAO. The categories are owners (*próprias*), renters (*arrendados*), sharecroppers (*parceiros*) or occupiers (*ocupantes*). The definitions of these are fairly broad; the first includes usufruct rights, for example, while renters can pay in produce, making them not dissimilar to sharecroppers. The last category, occupiers, is especially loose, defined simply as one who is occupying the land, with or without the owner's consent. There was also one interviewee who owned his farm, but rented about as much land again; however, this did not seem common.

The high percentage of owners is encouraging, because they would have a greater appreciation of any co-benefits produced by a sinks project. However, there is one category that may not be properly reflected in either the sample or the official figures; the occupier category in the sample includes an employed farm minder in the Itaperuna area.

This is apparently common in Rio and São Paulo states, where wealthy urban professionals – often lawyers or doctors – from the two conurbations apparently invest in land rather than other forms of property as it carries a lower tax burden, provided it is worked for agriculture. To satisfy this condition, a farm labourer may be employed to supervise the farm, but will have little technical knowledge and no authority to invest in new practices. A carbon project could not easily work with them. Asked how many farms in the area were managed in this way, a local extension agent thought it might be as high as 30–40 per cent; but most information about this practice is anecdotal.

Farm size

Farm size, like status, is broadly comparable with the region. Farm size of the sample ranged from 1.0 to 94ha and had a mean of just over 23.8ha. It is untypical of South-East Brazil only in the sense that there were no units of under 1ha or over 100ha; however, as the histogram suggests (Figure 8.2), the vast majority of farms in the four states fall within the same range. Moreover the mean, at 23.8ha, was not far off that for family farms in Brazil, which is 26ha (INCRA/FAO, 2000, p24).

Besides South-East Brazil, the farm sizes in the sample were also compared with the average farm sizes for Minas Gerais and Rio de Janeiro States. In the case of Minas Gerais, the sample appears to have overemphasized establishments between 1 and 5ha; but Minas is the size of

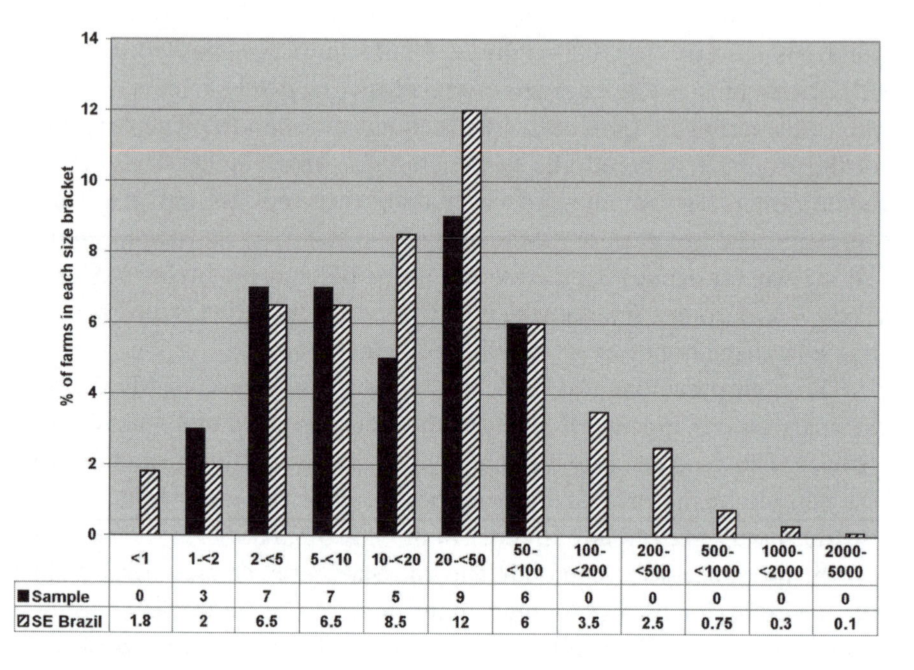

	<1	1-<2	2-<5	5-<10	10-<20	20-<50	50-<100	100-<200	200-<500	500-<1000	1000-<2000	2000-5000
■ Sample	0	3	7	7	5	9	6	0	0	0	0	0
▨ SE Brazil	1.8	2	6.5	6.5	8.5	12	6	3.5	2.5	0.75	0.3	0.1

Figure 8.2 *Farm size, sample compared with South-East Brazil*

Source: sample data; IBGE (1996)
Note: Percentages are approximate for sample data. IBGE data are rounded up/down to nearest 0.5%. IBGE (1996) reports just under 1.7 million farms in the four states, including 523 that exceeded 5000ha; these have been omitted from the histogram, but were incuded when calculating the percentages.

France, and there is bound to be considerable variation between agroecological zones. In Rio de Janeiro State, the sample shows a bias in favour of farms between 20 and 100ha, and against the small plots that would be typical of peri-urban agriculture elsewhere in the state (although these are represented to some extent by the INCRA settlement). In general, however, the sample farms were broadly typical of South-East Brazil.

Constraints faced by farmers

The results begin with some general information about the constraints perceived by farmers, as these provide important context for their reaction to the different proposed management practices. Farmers were asked about the chief constraints they faced and asked to rate them on a scale of 1 to 10. The constraints discussed here are general ones to the farming system. In particular, insufficient labour supply, especially in Noroeste Fluminense, would have a bearing on diversification of agriculture, tying farmers to relatively extensive, low-input practices such as pasture, while a shortage of credit, especially in the Zona da Mata, would restrict the use of inputs such as fertilizer.

Figure 8.3 gives farmers' ratings for the constraints. Although a useful guide, the results should be approached with caution, as the differences between individual ratings were sometimes statistically significant, sometimes not. For example, credit was clearly a more serious constraint than erosion, but the difference between credit and labour was not significant. For similar reasons, the data are not presented as disaggregated between survey sites; there were some apparent differences, and these are discussed if relevant, but the sample sizes limit their statistical validity.

Nonetheless, the high rating for credit is worth noting. At first sight it is puzzling, as the *Programa Nacional de Fortalecimento da Agricultura Familiar* (PRONAF), a federal organization that exists to support family agriculture, could advance up to Rs6000 (just over £1400). However, Banco do Brasil was perceived as less helpful, and PRONAF seemed to require considerable documentation, including two *avalistas* – guarantors – plus a house/flat as security. Buainan et al (2003, p343) confirm some

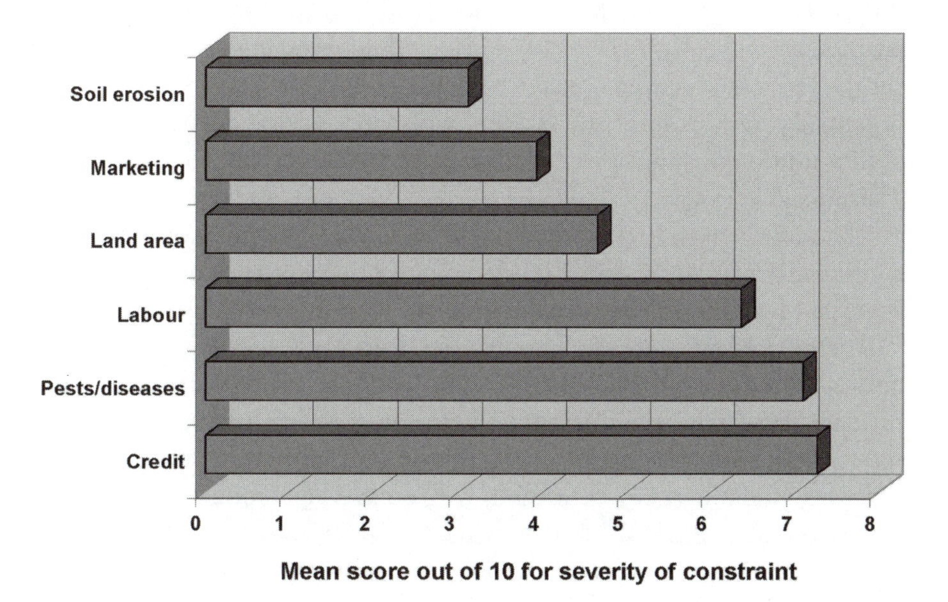

Figure 8.3 *Farmers' rating of constraints*

of this and identify a need to cover the extra transaction costs for family farms, provide coverage for risk and to resolve the problem of guarantees; they also call for a 'debureaucratization' of the regulations covering savings and collective credit. Until this is done, it is harder for farmers to invest in those measures they do want to take, such as seeds and labour for pasture improvement. According to the Empresa de Assistência Técnica e Extensão Rural (EMATER) extension agent in Piraúba (Zona da Mata), the lack of credit has led to a form of sharecropping whereby a farmer, being able to afford nothing more than the land rent, relies on others to provide absolutely everything else; they, not the farmer, then get the bulk of production. Under this system those who provide the bulk of on-farm investment would have no incentive to invest in sustainability, even if the farmer thought it necessary.

This may be subject to change as credit arrangements for farmers are brought into line with a sustainability agenda. In November 2010 the Brazilian federal development bank, *Banco Nacional de Desenvolvimento Econômico e Social* (BNDES), announced the *Programa para Redução da Emissão de Gases de Efeito Estufa na Agricultura*, an Rs1 billion programme under which farmers were to be lent funds for activities that reduce GHG emissions. Under the scheme, BNDES is willing to finance up to 100 per cent of the activity at an interest rate of 5.5 per cent (BNDES, 2010). The eligible initiatives include crop-livestock integration, silvopastoral activities, soil conservation (including terracing), and – crucially for the farmers surveyed – pasture improvement. The programme sounds potentially very helpful for farmers who wish to diversify for sustainability (although subject to constraints other than credit, such as labour – discussed below).

The second most highly rated constraint (although the margin was not significant) was pests and diseases. Tickfly and botfly, ants and, in the Zona da Mata, pests and diseases of tomato were amongst those identified.

The next constraint was labour – the third highest rated, but the gap between credit, pests and diseases and labour was not statistically significant. However, the difference between the two main sites in their rating for labour was significant. In Noroeste Fluminense, most farmers regarded labour as a key constraint, with an average rating of 7.8. For the Zona da Mata it was just 4.5. In fact, the Noroeste Fluminense farmers had just under 2.2 people working on the farm with an average farm size of just over 30ha, an average of 14.1ha per unit of labour. The Zona da Mata farmers, with their smaller farm size of 19ha, had an average of 3.8 people

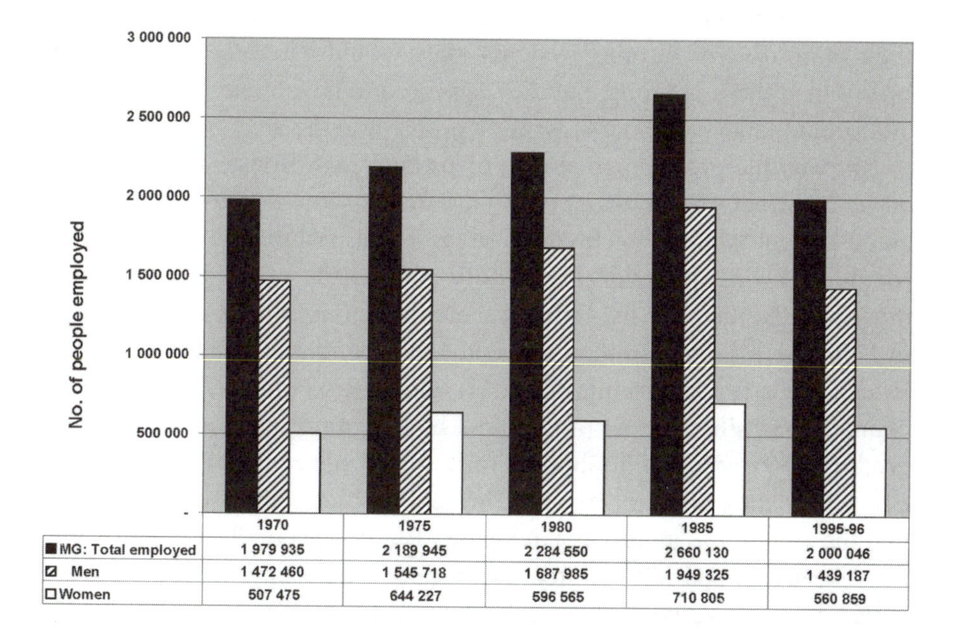

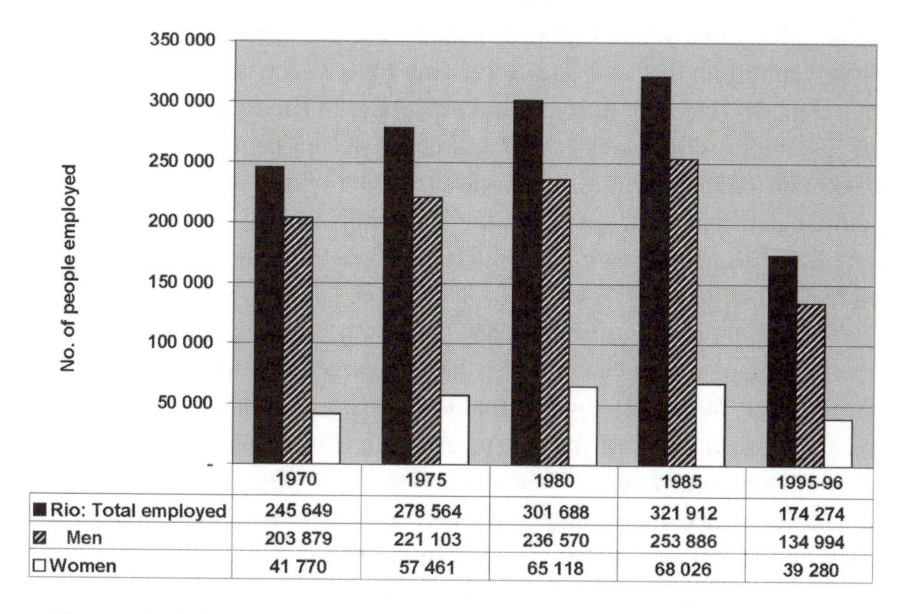

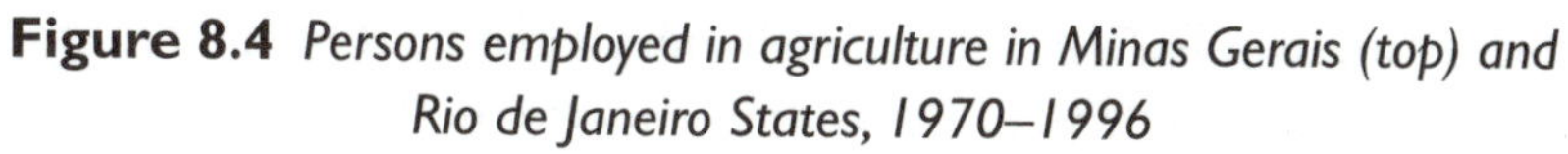
Figure 8.4 *Persons employed in agriculture in Minas Gerais (top) and Rio de Janeiro States, 1970–1996*

Source: IBGE (1996)

working on the farm, giving an average of just over 5ha per worker. With their more diverse farming systems, they would naturally require more labour than the first group, but they appeared to have more family labour, and could manage. Labour was still a problem in the area, however.

But was the greater percentage of pasture in Noroeste Fluminense a cause or effect of lack of labour? Farmers' responses implied that they could not get labour, which would suggest that pasture had been adopted for that reason. However, the history and profile of the area suggests otherwise. Neither did the farmers seem that anxious to obtain labour, as in Noroeste Fluminense they were paying around the minimum wage in Brazil. The artificial insemination (AI) staff in Rio Pomba (Zona da Mata) claimed that, since the AI programme had begun, farms would typically produce 150 instead of 50 litres of milk; so people *could* pay for employees, which they couldn't before. The real problem, they suggested, was that the labour was simply not available. But it was not clear whether this was because it had left the area, or because wages in other sectors were better. This is discussed again in the next chapter.

Whatever the reason, agricultural employment in Brazil has been declining for some years. Figure 8.4 shows the number of people employed in agriculture in the two states according to five successive surveys. There is a clear decline in both states, but especially in Rio de Janeiro, with a fall of just under 47 per cent (men) and about 42 per cent (women) between 1985 and 1995. In Minas there was a decline of about 26 per cent (men) and 21 per cent (women) over the same period. This takes the number employed in Minas back to about 1970 levels; and in the case of Rio, to rather below them.

The numbers should be approached with caution, not least because the hectarage worked would also have changed. Moreover Helfand and Brunstein (2000, p14) suggest that the decline of labour in agriculture is only a third to a half of the census estimate (ibid.). However, as so many farmers mentioned labour as a constraint, especially in Rio, these figures are probably not misleading; and where labour is short, extensive systems may be the only ones that can be improved for carbon content, even if other options are theoretically superior. This increases the importance of pasture, with its relatively low labour requirement, in this survey.

The next constraint (although some way behind in importance) was land area. Here the situation is reversed; the Minas farmers would have liked more land, but the Rio farmers were less concerned. This may reflect the more intensive land use in Minas, but it might also have been expected that the Rio farmers would want larger areas to compensate for the poor fertility of the land – a continuation of the extensive, extractive model. It may be that they themselves now see the fertility of existing land, rather than its extent, as the issue, and their enthusiasm for pasture improvement may imply that.

Below came marketing. The farmers were asked to rate marketing on the assumption that it covered the profitability of the product as well as its ease of sale, but sometimes took it as only the latter. With this they had no problem, as there was a large milk-processing factory in Itaperuna. Whether prices were adequate was another question. It was claimed that they received just 70 centavos (just over 17p) a litre; but milk was being sold in a supermarket near Rio city for Rs1.49, more than 35p. Two Noroeste Fluminense farmers added that, while they could sell milk easily enough – albeit at a low price – other products such as fruit were harder to sell. One farmer had suspended attempts to produce fruit commercially because the market was poor. However, had farmers attempted to sell as a group, a buyer might have been found. With the Zona da Mata farmers, it was made clearer that marketing included all aspects of sale and this resulted in a higher rating (6.1). They were even more disgruntled by their prices. One commented that the intermediaries who brought their produce were 'sucking their blood'. Farmers, it is true, rarely complain of being overpaid. However, the EMATER agent in Piraúba did concur that much of the profit was taken in this way. He suggested that farmers needed to get together in groups in order to market their produce themselves.

Soil erosion came next. It was only slightly lower than marketing. But it *was* rated significantly lower than all the other constraints. A third of the respondents actually gave a rating of zero. Yet erosion was clearly seen on some farms; and when driving up the Muriaé Valley from Campos dos Goitacazes, it was all too evident. This may reflect the fact that an area of erosion has less impact in an extensive system where the cattle can always graze elsewhere. It may also indicate that farmers had problems they considered more significant. This will limit the extent to which some 'carbon-friendly' practices are welcomed – as discussed in the previous section, farmers will not always wish to adopt erosion control if the erosion is not causing them any problems, and the fact that it is emitting soil carbon will be of little interest to them. One or two farmers did rate erosion as a serious problem, but they were observed to have steeply sloping land.

The low rating for erosion is important as its prevention is a key to reducing emissions from agricultural land. The obvious implication is that an agricultural sink is not in this case congruent with farmers' priorities. However, the picture is more complicated than that, for three reasons. First, although few farmers rated erosion highly as a problem for them, two remarked that it was for other people. One farmer in the Zona da Mata gave a rating of 3 but said that it would be 7 for the area. It is possible that not all those affected wanted to admit that the problem was serious on their land. Second, some farmers who did not rate erosion especially high, did comment on other soil problems such as soil fertility and acidification. However, they did not connect that with the question on erosion; and indeed they were not meant to. It might be that had the constraint been framed as 'soil quality, including erosion', the mean would have been much higher. Given that nutrients can be lost through runoff whether erosion is apparent or not, that would in retrospect have been one way to frame the question. And third, practices that reduce erosion might be positive for farmers in other ways, so the fact that erosion was not seen as a priority does not mean that combating it for carbon purposes is not practical. Nonetheless, there is a warning here that it will not automatically be seen as beneficial by farmers, and indeed will not always deliver concrete benefits to them.

The farmers were not always asked whether access to technical information was a constraint, as the presence and role of the extension officer at some of the Rio sites would have distorted the responses. (This obviously did not mean that he was *not* giving good advice). However, in some locations farmers were critical of the state extension organization, EMATER. In fact, the extension agent based in Itaperuna complained that he was badly under-resourced.

The Minas farmers also stated that they sometimes lacked information, especially about fertilizer rates and application. They were less directly critical of EMATER, but one pointed out that there was only one extension agent to cover the region; that agent was helpful when interviewed, but believed that the farmers locally were backward and were always wasting fertilizer. This seemed a little unfair since at least two of the farmers interviewed that morning had expressed a need for information on exactly that point. It seems fair to conclude that access to technical knowledge was a constraint.

Rating the Practices

The following section reviews farmers' responses to the questions about carbon-friendly practices. Not all are given equal space. Pasture improvement is clearly key in this farming system; because of its close relationship to inorganic fertilizer, the latter is almost as important, and the two relevant subsections should be considered together. Organic agriculture is given some space; its potential to mitigate climate change is controversial, but it addresses the needs of some farmers and could have other environmental benefits. Tillage regimes and associated rotations are also important. Some other options did not prove so relevant in practice and have been discussed in less detail.

Pasture improvement

The extensive nature of Brazilian agriculture is revealed by the percentage of its agriculture that is pasture. Nationally, it accounts for about 50.25 per cent of farm area; of this, about 22 per cent of total farm area is natural pasture and 28 per cent improved. Boddey et al (2006, p312) point out that real natural pasture is very rare. All 'natural' means in the IBGE definition is that the farmers did not plant the pasture themselves and do not believe anyone else did. In fact, it is the result of forest clearance, perhaps for pasture but more probably for coffee. This 'native' pasture is covered by the fragrant molasses grass, *capim-gordura* (*Melinis minutiflora*), *jaraguá* (*Hyparrhenia rufa*) and *colonião* (*Panicum maximum*, or guinea grass). These came from Africa, apparently as bedding in slave ships, and were spread by animals and the wind so that they colonized deforested areas (Boddey et al, 2006, p312).

In Noroeste Fluminense, the percentage of agricultural area that was pasture in the sample is broadly similar to the local norm (80.5 per cent against 77.2 per cent). Seropédica is anomalous – only one of the farmers in the INCRA settlement had pasture; however, peri-urban pasture can be important, and a 2002 socio-economic survey of Seropédica suggests that pasture makes up 44.5 per cent of the sub-municipality, and 82.5 per cent of the agricultural land area (Tribunal de Contas do Estado do Rio de Janeiro, 2002, p12). As for the Minas Gerais sample, the percentage of pasture appeared to be a little below the local average, but this may have reflected very local circumstances – perhaps even the way one large landholding broke up after the coffee failure. In general, sloping land was being used for pasture, while the flatter valley bottoms were given over to more profitable use. In general, Boddey et al (2006, p312) suggest that pasture is slightly over 50 per cent of the land area of the Atlantic Forest biome.

It may be assumed that pasture is an inherently unsustainable land use compared with the native forest that once occupied the land. But as far as soil carbon is concerned, this is not so. Where forest has been converted directly to pasture, the SOC level (as distinct from above-ground carbon) may actually rise. Boddey et al (1997, p789) cite reports of SOC increasing from 35 to 63t/ha in the 0–30cm layer eight years after pasture establishment in succession to native forest in the Manaus area. Fisher and Thomas (2004, pp123–124) quote data from Rondônia that suggest increases of 16t/ha over a rather longer period. This is not to argue that pasture is somehow preferable to forest in terms of carbon storage, as the total system carbon including above-ground carbon is clearly higher in a forest. Nonetheless, SOC content is linked to the soil organic matter and thus to soil structure and plant growth, so these figures imply that pasture can be a sustainable long-term land use *and* carbon sink if well managed.

But it is often not. Boddey et al (2006, pp311–312) report that the 20.5 million ha of 'natural' pasture in the biome, mostly on hillsides, must 'almost entirely' be classified as degraded. There is about the same amount again of pasture that is planted, mostly with *Brachiaria* spp., but Boddey et al state that 'much' of this is degraded as well; they ascribe this to overgrazing and lack of chemical fertilization (Boddey et al, 2006, p312). Boddey et al (1997, p790) say that nitrogen immobilization through large quantities of plant residue deficient in nitrogen is also a factor, and suggest that incorporation of forage legumes would also be a good strategy.

There is little data from the Atlantic Forest biome on how much SOC might be sequestered from improvement of the 50 million ha of pasture in the region, but what data do exist are encouraging. Tarré et al (2001, p18) reported a 43 per cent increase in SOC in the top 5cm over a six-year period and 22 per cent in the 5–10cm layer over nine years. They found that, on a grazed pasture sown with *Brachiaria*, SOC accumulated by 0.66Mg/ha/yr C to a depth of 100cm. Incorporation of a forage legume improved this to 1.17Mg/ha/yr C. The pastures had been fertilized at low rates with phosphorus and potassium but no nitrogen. This experiment was in Bahia, some way to the north of the study area but within the biome.

There is no local tradition of incorporating forage legumes. There is a strong argument for them in tropical pastures, given that nitrogen is normally a constraint. Giller (2001, p187) argues that since fertilizers tend to be expensive and unavailable for this purpose, incorporating legumes is the only option if pasture productivity is to be improved. However, Boddey et al (2003, p610) record that early experiments with this were not very successful, due to low persistence of introduced plants that had performed well in Australia. In recent years there has been some success in identifying varieties of (for example) *Stylosanthes* that persist in sward with *Brachiaria* (ibid.). There might also be potential benefits for the farmer in terms of liveweight gain from properly managed mixed swards in the region (Cóser et al, 1997, p18).

But there are constraints to adoption, not least the need for fertilizer for establishment and pre-treatment to ensure uniform germination (Giller, 2001, p200). In the survey, the emphasis was on improvement with *Brachiaria* alone, and the farmers were asked about pasture improvement in the sense in which they would normally understand it.

This involves ploughing pasture and sowing with *Brachiaria* spp., not necessarily with much in the way of inputs (as described later, fertilizer cost is seen as a serious constraint). But this is still profitable for the farmer, and potentially worthwhile for carbon sequestration. In an experiment very close to the Zona da Mata site in the late 1980s, Cóser et al (1997, pp17–18) recorded liveweight gains of 110.8kg/ha on *capim-gordura* (the fragrant molasses grass, *Melinins minutiflora*), one of the 'native' grasses, but 177.8kg/ha on *Brachiaria*. Moreover the latter establishes itself well on compacted soil. Fisher and Thomas (2004, p128), writing on tropical lowland pastures in South and Central America in general, comment: 'There are a number of data for well-managed pastures of introduced grasses of African origin that show accumulation of C in the soil at rates close to 3 t/ha.' Although they think this figure can be extrapolated safely to areas sown to *Brachiaria* species in the eastern plains of Colombia, they are less sure about pasture in the Brazilian Cerrado and Amazon, and do not discuss the Atlantic Forest biome. Moreover, their focus is on regions that have been converted from native forest or grassland in the last 30 years; as discussed in the previous chapter, the pasture in the survey area succeeded other land uses, and had ceased to be native forest much earlier. Nonetheless this figure is food for thought. Could this potential be harnessed with the farmers in the survey?

They were asked three questions about pasture. First, if the farmer was practising pasture improvement, how pleased was s/he with the results? Second, if the farmer was not doing so, how much did they want to? And third, how good did they think this practice was for the soil? (Soil health, here, is a proxy for organic-matter content and therefore carbon.) The third question asked of farmers was asked of both those using the practice and those who were not. This technique was followed for all the practices reviewed.

Box 8.1 A flexible farmer?

Although there was pasture in the Seropédica area, only one of the farmers surveyed had any. He had a 4ha farm of which about half was pasture, on which he kept four cows, two bulls and a heifer; he grew okra on most of the rest. He owned the pasture, but said that the land on which he was growing okra was sharecropped, with 20 per cent of the crop going to the owner. Later in the interview he said that the owner of the land was INCRA; it was not clear if it was the ultimate owner of the land, or if he was sharecropping from a tenant. Although okra and pasture between them were his main living, he had planted rice and corn in the past, so might be ready to change his system again. He practised rotation. A user of inorganic manure, he had acid soil and thought this a serious problem – but he did not know anything about green manures. He was a farmer who might have had the flexibility and incentive to move to carbon-friendly practices if assisted.

The responses showed that pasture improvement was very popular with farmers. About one in three were using it; those who were not using it wanted to, and those who already were wanted to use more. Those who answered the first question were fairly consistent, rating pasture improvement at 7 to 10 out of 10. Although the Zona da Mata farmers were less dependent on pasture, they gave a response that was only slightly lower than the overall mean. They did place a little less emphasis on pasture improvement being good for the soil (7.9 against 8.7), which may reflect their more negative feelings about inorganic fertilizer, discussed later. However, the difference was not statistically significant.

Figure 8.5 gives farmers' average rating out of 10 for different constraints to pasture improvement. Some farmers did not give a response, but usually stated that they lacked the knowledge to give an opinion on a question, so there is no reason to suppose that excluding them from the mean introduces a bias.

Figure 8.5 seems to reflect the overall constraints to agriculture expressed earlier (credit and labour). This is probably because pasture was, for many, the key activity, so when asked about constraints to farming, they identified those that constrained it. The high rating for initial cost would reflect tractor hire as well as fertilizer – and perhaps also labour, which was asked about separately and was also rated high. These ratings are understandable because, if it is done properly, improving degraded pasture with *Brachiaria* is not necessarily cheap, even if there is no attempt to use forage legumes. Boddey et al (2003, p608) quote estimates (from the Cerrado region) of US$350–400/ha, involving decompaction through ploughing, seedbed preparation, fertilization and reseeding.

If cost and labour are the main constraints, then external funding linked to carbon sequestration might address them well, helping farmers to do something that they already want, and know how, to do. The tonnage of carbon could be worthwhile, too; that is discussed further in the next section, when fertilizers are rated. Two qualifications should be made, however.

First, if forage legumes were used, there would be more benefit in terms of carbon sequestration – but farmers were not familiar with this technology. Had they been asked about it, they might have rated technical knowledge higher as a constraint.

Second, there might be little point in trying to sequester carbon through pasture improvement if this is then counterbalanced by farmers increasing stocking rates. This is complicated, however, as carbon sequestration and sustainability can sometimes be improved by *increasing* stocking rates. Muchagata and Brown (2003, p807) point out that grass growth can – up to a point – be stimulated by consumption; if stocking rates are too low, 'a significant amount of biomass is produced, [but] the weed invasion is considerable, and the risks from fire high'. They argue that well-managed, appropriate intensification of cattle production on existing pasture in the Amazonian basin could reduce pressure on the forest frontier.

This is not so far from the arguments described in an earlier chapter – that intensification reduces deforestation; but it will be remembered that there were considerable qualifications to that argument. Moreover, higher stocking rates would carry a risk of increased CH_4 emissions from the

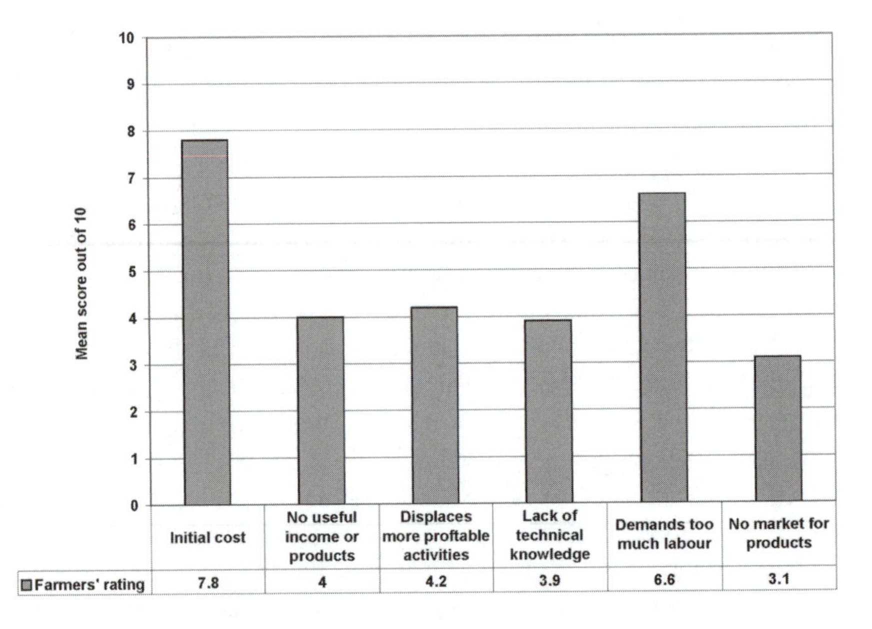

Figure 8.5 *Farmers ratings of constraints to pasture improvement by ploughing/seeding with* Brachiaria

cattle themselves. These increased emissions can be quantified, and to some extent can also be mitigated (for a discussion of these issues, see for example Kelliher and Clark, 2010). However, in other respects, the results of such higher stocking rates are not predictable.

Boddey et al (2004) describe an experiment in Bahia in the north of the biome in which stocking rates were increased from two animals per hectare (which would be higher than normal for the survey sample here), to four animals per hectare. Liveweight gain per animal decreased, but overall liveweight gain showed some improvement. However, the way in which nitrogen was cycled through the system changed considerably; although the total exported from the pasture as weight gain did not change much, the amount deposited as litter decreased, that in urine and dung increased and the latter was often deposited around drinking troughs or in other areas where it was too concentrated, or the grass too trampled, for it to be taken up. Given that nitrogen deficiency is a key to pasture degradation in the region, this is clearly not helpful. Moreover, while Boddey et al were mainly concerned with pasture sustainability, there are also climate-change implications, as nitrogen volatilized in dung or urine rather than taken up will cause emissions of N_2O, a powerful greenhouse gas.

For the sake of this survey, then, it is assumed that the result of pasture improvement on stocking rates and vice versa is an unknown, but it would clearly need to be considered as part of any project.

Inorganic fertilizer

As has been discussed, underuse of fertilizer (inorganic or organic) is a major cause of loss of organic matter in the region's pastures, and any attempt to increase soil carbon would have to confront this. Farmers' attitudes to fertilizer – whether they use it, or would like to use more, and what prevents them doing so – are therefore important.

Fertilizer has been an essential component in improving agricultural productivity. Fresco (2003, p4) reports that in 1950, farmers worldwide applied only 17 million tons of mineral fertilizers: '[F]our times more than in 1900 but eight times less than today.' The increases in yield as a result are quite startling; in France, for example, wheat yields have increased from about 1800kg/ha in the 1960s to over 7000kg/ha today (Fresco, 2003, p4). However, the emphasis of the survey question was on the use of fertilizer for pasture, as that is its main use in the survey areas, and would have much the greatest impact on soil carbon, as the hectarage of other land uses was relatively small.

But using fertilizer in an extensive farming system such as the Atlantic Forest pastures will give less return per hectare than it would with a high-value crop. Thus one or two farmers used fertilizer on small areas of sugar cane, which is sometimes ground up for feed; and the Zona da Mata farmers also used it for tomatoes, with a residual benefit for maize where tomatoes were rotated with it. But for pasture, farmers might have a harder decision. Yet they may still want to use it, and it may be just as important. As observed earlier, adequate cycling of nitrogen through the system is crucial to organic matter production. So farmers are caught in a paradox; fertilizer is essential in a system where relatively few nutrients are being cycled back over a wide area, and alternatives such as crop residues and legumes may be problematic. Yet it is not always profitable to use fertilizers.

There is also a climate-change cost. Nitrogen fertilizers, livestock production and gaseous losses from manure between them account for 60 per cent of emissions of N_2O (Norse, 2003, pp2–3), a greenhouse gas with much greater radiative forcing potential than CO_2.

It should be possible to quantify this. IPCC methodology states that for every 100kg/ha of nitrogen applied as fertilizer, the global warming potential (GWP) is equivalent to 450kg/ha of CO_2 (Robertson and Grace, 2004, p55). Over a 20-year period, N_2O has a radiative forcing potential 275 times that of CO_2, and 'an agronomic activity that reduces N_2O emissions by 1 kg ha^{-1} is equivalent to an activity that sequesters 275 kg ha^{-1} CO_2 as soil C' (op. cit., p53).

> *Rates of nitrogen application vary widely, largely as a function of market availability, crop value, and national subsidies; typical rates in developed regions range from 50 kg Nha^{-1} for wheat to 200 kg Nha^{-1} for maize. For every 100 kg Nha^{-1} that is applied, the GWP cost is 45 g CO_2-equivalents m^{-2}.* (Robertson and Grace, 2004, p55)

A rate of 45g/m^2 is equivalent to 450kg/ha, or 0.45t/ha. Application rates on pasture in the biome will be near the lower bound of the rates quoted above, however. In the experiment of Cóser et al (1997) in the Zona da Mata, described in the previous subsection, 40kg/ha of nitrogen was applied in December 1985, then from 1988, 50kg/ha every two years, or an average of 25kg/ha per year. This would imply emissions equivalent to roughly 11kg (0.11 of a ton) per hectare. If Fisher and Thomas (2004, p127) are right and a pasture improved with *Brachiaria* or equivalent can sequester SOC at rates approaching 3t/ha C, which is quite a high figure (see previous section), there is a considerable net reduction in atmospheric CO_2 even after allowing for emissions from fertilizer manufacture, transport and application.

These calculations are extremely crude. They take no account of the wide variation that would occur between farmers' use of the correct application rates, the soil type or other factors, all of which would have a bearing on the real rate of nitrogen volatilization. Also, as described in the previous subsection, Boddey et al (2004) established that stocking rates on pasture have an enormous effect on the way nitrogen is cycled through the system, and on how much is lost from dung and urine. It must also be stressed that response to fertilizer is at least partly dependent on the presence of organic matter, and there is a limit to what chemicals can do without organic inputs; this is a subject raised again later in this chapter.

However, these figures, however crude, do suggest that *net* CO_2 sequestration in pasture is practical using nitrogen fertilizer – a point which could reasonably have been questioned, given the forcing potential of

N_2O. Moreover, fertilizer on pasture is probably more congruent with farmers' other goals and realities, offering higher earnings per hectare from cattle rather than from other strategies that are labour-intensive and therefore sometimes impractical in the region.

This was reflected in farmers' responses. Asked if they liked inorganic fertilizer, they gave an average rating of 8.3 out of 10; asked if they would like to use more, they gave an average rating of 6.6 (this figure would be lowered by the responses of the Zona da Mata farmers, who were more cautious of it). Asked how good they thought it was for the soil, the farmers gave a rating of 6.3 out of 10. Again, lower ratings came from the Zona da Mata farmers – 4.6 out of 10, against 7.6 for the rest of the sample – a significant difference. Given that the Noroeste Fluminense farmers rated fertilizer so much higher for soil health, this would clearly be a popular sequestration tactic for those with extensive livestock systems.

It is therefore important to know why farmers are apparently unable to use fertilizer as much as they would like to. In fact, the main constraint to using the practice was simply money; farmers were almost universal in their complaint that fertilizer cost too much. (True, farmers rarely complain that inputs are too *cheap*.) Lack of technical knowledge also stands out, but the difference between the two was statistically significant; farmers clearly thought the cost more important.

The next constraint was labour, but the rating was not significantly lower. The remaining constraints, 'No useful income', 'Displaces more profitable activities' and 'No market for products', were marked rather higher by the Zona da Mata farmers than by the others. Although this was not statistically significant, conversations with farmers suggested that it *was* significant in practice. The differing ratings of the Zona da Mata farmers are discussed below.

The survey seems to indicate that, first, farmers are keen to use more inorganic fertilizer and, second, that the main constraint to its use is initial cost. It might therefore seem that the best way to sequester carbon in the biome is to provide cheap and easy credit – or even subsidies – for fertilizer purchase.

However, the ratings, and remarks made by farmers, only partially bear this out. One reason is the impact of chemical fertilizers on the long-term health of the soil and on the broader environment. In Noroeste Fluminense, farmers did not appear particularly worried about these. In the Zona da Mata, however, just over half the sample expressed concerns about long-term soil health – even though they were not asked about it directly. Their main concern was acidification of the soil, but a few also expressed concerns about environmental pollution. For the time being, however, they had insufficient information about alternatives to chemical fertilizer (see Box 8.2). These concerns are further reflected in the discussion of organic agriculture in the next subsection.

Box 8.2 A farmer questions fertilizer

One of the Zona da Mata farmers who expressed doubts about fertilizer was farming 14ha, of which about half was pasture; as usual, the latter was on the more sloping land. On the remainder, he had about 4ha of maize, some tomato and smaller amounts of fruit and vegetable, and had planted 0.5ha of eucalypts. The farm was run with family labour, and he looked after the cultivated land while his semi-retired father looked after the pasture and their small dairy herd (20 head). Fertilizer, he said, was good for plants but not soil. Like others locally, he rotated so that maize got residual benefit from the fertilizer used on the tomatoes, but they lacked information on how much to apply, and he thought they might be using too much. But they had to use it because there was no alternative; when an organic fertilizer was produced locally some years earlier, he had been too sceptical to try it. A local company was apparently making another, but information was lacking.

It is not clear exactly why the farmers in the Zona da Mata were more concerned about these points. It is possible that the Zona da Mata farmers were influenced by proximity to the Rio Pomba campus of the Centro Federal de Educação Tecnológica (CEFET), which has courses at Rio Pomba that cover agroecology and sustainability. Their priorities may also be partly set by the Instituto Mineiro de Agropecuária (IMA), which oversees farmers' treatment of the environment in Minas Gerais.

The farmer quoted in Box 8.2 was fairly typical of the Zona da Mata; comments from others on chemical fertilizer included 'bad for soil – have to stop using it after a few years', 'sometimes they don't work very well and I have no one to ask why', and 'chemicals frighten me – I don't know if they are poisoning the soil'. One Zona da Mata farmer said that the publicity about agrochemicals in the media meant that they might soon have problems selling anything produced with chemical fertilizers. Another said he could have problems selling products if he made mistakes with fertilizer. Two or three of the Zona da Mata farmers also highlighted lack of advice on how best to use fertilizers and in what quantities.

Funding for inorganic fertilizer should be part of any carbon sequestration project for a region of this type. It had better carbon returns than most practices, was not especially constrained by the labour shortage, and, in the case of Noroeste Fluminense, it was what the farmers wanted and knew they could do.

However, such funding should come with adequate technical advice. It might also be wise to make it part of a package for pasture improvement that would include other measures, possibly sowing of forage legumes (mentioned above, although any package would have to be carefully designed). Another intervention that could be coupled with fertilizer is the use of leguminous tree species, but these are covered later.

Moreover, there will be farmers or groups of farmers who, like the Zona da Mata farmers, would prefer to look at alternatives to chemical fertilizer. Organic agriculture might be one of those.

Organic agriculture, green manure and pigeonpea

With its stress on maintaining organic matter, organic agriculture should, in theory, increase the carbon sink – but may also contribute to other environmental goals. Brazil's federal government has been broadly sympathetic to organic agriculture, and it has been expanding for some years; Neves et al (2000, p11) report that the market for organic products in Brazil expanded by 40 per cent a year in the late 1990s.

There is also a threat to public health from agrotoxins in the water supply in much of Brazil. According to IBGE (2002, p92), 43 per cent of Brazil's municipalities reported a problem with agrotoxins, including fertilizer as well as pesticides, and water pollution; in the South-East region, which includes the survey areas, it was 35 per cent.

Organic agriculture may reduce pollutants, but whether or not it mitigates climate change is an area in which strong opinions are sometimes held. In theory, it should have inherent mitigation potential because of the link with organic matter. Morison et al (2005, p25) say organic agriculture may be defined as a system 'that seeks to promote and enhance ecosystem health while minimising adverse effects on natural resources'. So, by definition, organic agriculture should conserve organic matter and thus soil carbon. The IFOAM definition of organic agriculture adopted in September 2005 affirms the 'ecological principle', which:

> *roots organic agriculture within living ecological systems. It states that production is to be based on ecological processes, and recycling. Nourishment and well-being are achieved through the ecology of the specific production environment. For example, in the case of crops this is the living soil …* (IFOAM, 2005)

In a paper published by the International Federation of Organic Agriculture Movements (IFOAM), Kotschi and Müller-Sämann (2004, p18) say that organic agriculture 'is a systematic strategy, which may reduce GHG emissions and may enhance the sequestration of carbon, the most important green house gas'. They state that reliance on and quantity of nitrogen fertilizer have become the main reasons for reduced SOM (op. cit., p19).

But one could also argue that more is lost following *insufficient* use of nitrogen fertilizer. Equally difficult is their contention that organic agriculture can prevent emissions through deforestation because of its sustained productivity (op. cit., p21). As mentioned in previous chapters, the suggestion that maintenance of soil fertility might reduce deforestation has also been used by Vlek et al (2004) to advocate *greater* use of fertilizer. Cowie et al (2007a, p340) put things the other way round, suggesting that lower yields under organic agriculture could cause leakage by inducing conversion of additional land to arable.

The best way to look at organics and climate mitigation is to compare fertilizer with organic inputs. The latter includes compost, which might be made with both animal and plant material. In general, however, organic inputs can be split into two basic types: farmyard manure (animal excrement) and green manures, which could be any crop grown mainly for soil fertility.

Edmeades (2003) reviews results from 14 long-term (20–120 years) field trials that compared farmyard manure, slurry and green manure on crop production and soil properties.[2] The effect of manures on yields does not seem to be any greater than that of equivalent amounts of fertilizer (Edmeades, 2003, pp177–178). However, manure did generally seem more effective in increasing soil organic matter. As Edmeades points out, this is to be expected as it contains organic matter itself, and he does expect manures to have a yield advantage when used over many years so that there is a large build-up of organic matter. In fact, four of the farmers in Noroeste Fluminense reported that they used cattle manure and thought it very useful. However, one commented that it was difficult to use and another said that it was hard to apply on sloping land. In the Zona da Mata, only two farmers reported using cattle manure (although these were only the farmers that volunteered such information; it is likely more farmers did use it).

It does not entirely remove the problem of pollution. Edmeades (2003, p178) notes that application of manure may under certain circumstances cause soil phosphorus enrichment, leading to greater runoff; and that manures are more likely than fertilizers to add to poor water quality. Despite this, in the 2002 IBGE survey of environmental quality in municipalities, just 226 out of 5560 reported water pollution due to livestock production; in the south-eastern states, it was 75 out of 1668 (IBGE, 2002, p340). However, it may be that such pollution is only reported when it is very obvious, and that some water-quality issues are wrongly ascribed to agrochemicals.

Another alternative to chemical fertilizer is green manure. This could be broadly defined as a crop that is grown for the sake of soil fertility, although as Vandermeer (1995, p211) points out, it is also sometimes grown to control weeds or insects. Again, the key questions are, first, whether it can replace inorganic fertilizer for productivity, and second, whether it builds up sufficient organic matter to enhance the carbon sink.

Regarding the first question, comparisons are difficult because it is hard to compare like with like; for example green manure might succeed in matching productivity from a given land area, but with a far greater labour input. Thus Giller (2001, p177) quotes data from Malawi suggesting that turning the mulch from a stand of mucuna into the soil on 1ha takes 60 man-days. However, in the live-mulch system used in parts of Central America, where mucuna is intercropped with maize, the labour requirement actually drops because there is less need to weed (op. cit., p179). In any case, comparing the use of green manure to inorganic fertilizer assumes that the farmer had a choice, and as the previous subsection suggested, farmers in the survey area find fertilizer expensive. (It is worth noting that, while green manures may replace nitrogen fertilizer, inputs of phosphorus and potassium may still be needed.)

The second question – whether green manures can build up sufficient SOM to sequester significant amounts of CO_2 – is also hard to answer.

Green manures may not have long-term effects on SOM (or yields), because they decompose quickly (Giller, 2001, p178), but this may not always be the case if there are sufficient inputs. Green manure also has potential for increasing soil carbon if used to replace fallow; in the southern State of Rio Grande do Sul, it was found that a mucuna/maize rotation plot under no-till sequestered 15.5Mg/ha CO_2 over 8 years, compared to a net emission of 4.32Mg/ha CO_2 from the traditional maize/fallow plot (Evers and Agostini, 2001, p6) – but this would be explained at least in part by the change in tillage regime.

Box 8.3 Important to invest in soil

In Noroeste Fluminense, a middle-aged woman was farming just 4ha, three-quarters of it pasture, with 10 dairy cattle. The remaining 1ha provided guava, mango and numerous other fruits for home consumption. She had only had the farm a year, but had worked for 20 years before that on her grandfather's land. There were no employees, but she was helped by her teenage daughter. Both felt strongly that it was necessary to invest in the soil. Although they would have liked to use more inorganic fertilizer, they were aware that green manure would be better for the soil, and mentioned the need for fallow. The farmer had also heard of vegetative strips and approved of contour planting (by contrast, while she was being interviewed, a neighbour was ploughing straight up the steep hill opposite). She knew about pigeonpea, but not as a potential crop – it was, she said, good to eat with chicken. The EMATER agent had discussed organic agriculture with her and her neighbours, and she approved in principle. Unlike some farmers who thought they were organic because they used few chemical inputs, she was aware of the need for certification.

However, if a leguminous green manure were to increase available nitrogen, it might increase SOM indirectly by increasing productivity of a succeeding or intercropped crop, or indeed pasture – a possibility discussed with reference to pigeonpea later in this section. Moreover, if a green-manure rotation were to reduce chemical input, there would be a net saving in emissions even if it does not increase the carbon sink.

Indeed, Kotschi and Müller-Sämann (2004, p22) point to the energy demands of synthetic fertilizers as one of the reasons why organic agriculture has mitigation potential. But there is no gain from this if no such fertilizers were previously in use, so proving additionality for a carbon project would in some cases be hard. Moreover, as they themselves point out (op. cit., p19), although IFOAM guidelines prohibit 'Chilean nitrate and all synthetic nitrogenous fertilizers, including urea', they do not expressly exclude all mineral fertilizers. Where local organic certification guidelines permit this, the mitigation potential of organic agriculture in pasture could be enhanced; as Boddey et al (1997, p790) report, forage legumes can be introduced into pasture with modest amounts of lime, phosphorus and potassium fertilization (see also Tarré et al, 2001). Because they compete poorly with grasses for soil nitrogen, the forage legumes will obtain nitrogen through biological nitrogen fixation (BNF). As Boddey et al point out, this may be an alternative in the large areas of Latin America where nitrogen is often immobilized by nitrogen-deficient plant residues but nitrogen fertilization is not economic.

In the survey area, then, organic agriculture might indeed increase soil carbon relative to the probable alternative without intervention, which would be the continued use of chemical fertilizer but in inadequate quantities. What are the prospects for adoption?

Farmers were enthusiastic about organic agriculture, but there is an anomaly in the sense that some thought they were actually practising it, because they were not using inorganic fertilizers. However, they were not certified; some might have been using pesticides, and one of the Noroeste Fluminense farmers who said he was organic was actually using urea on sugar cane (which is used for fodder), although urea is debarred by most definitions of organic agriculture. This does not mean that the farmers' enthusiasm was meaningless – on the contrary, they were effectively saying that not using inorganic inputs worked for them. However, a more reliable response came from the question as to how much farmers would *like* to practise organic agriculture. A high mean of 8.6 is based on responses from about 75 per cent of the sample. This would suggest that willingness to go organic may not be unusual in the region. The most obvious reason is higher farm-gate prices for not using inputs that farmers cannot afford to use much anyway.

However, a third question – how good did they think the practice was for the soil – suggests another reason. Farmers gave a mean rating of nine out of 10. This suggests a widespread disenchantment with chemical inputs. In fact, two of the Noroeste Fluminense farmers explicitly stated that they did not think chemicals were good for the soil, while those in the

Zona da Mata were quite vocal; seven said that chemical fertilizers were bad for the soil, and another blamed it for blight on his tomatoes. In fact, the Zona da Mata farmers seemed to like organic agriculture not simply because organic agriculture was good for the soil in itself, but through disillusion with the alternative. Their comments bore this out.

However, there are barriers to adoption of organic agriculture. The farmers perceived the main one as being technical knowledge, but there was a significant difference between the ratings for Noroeste Fluminense (5.6) and the Zona da Mata (9.1). It was noted that the agent in Noroeste Fluminense had already discussed organic agriculture with the farmers at at least one location in the area, whereas his colleague in the Zona da Mata was quite negative about organic agriculture.

Regarding initial cost as a constraint, there was a wide spread of ratings. It may be that some farmers assumed that all they had to do was stop using inputs they often could not afford anyway, while others did realize that obtaining certification can be costly. But farmers might not always have to meet these costs in full; in over 35 per cent of municipalities, the authorities claim to offer incentives for organics (IBGE, 2002, p199), and

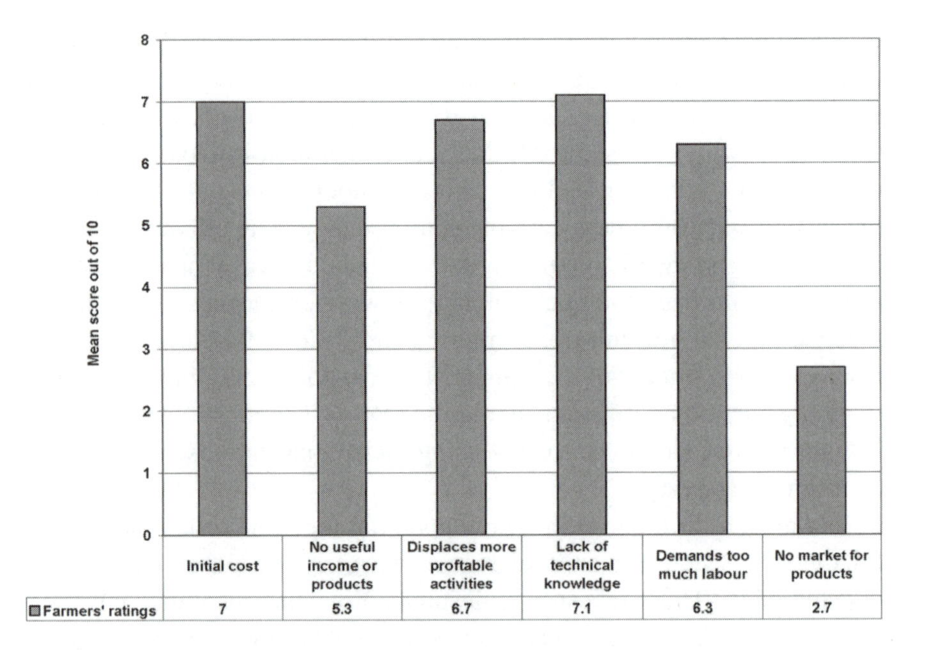

Figure 8.6 *Farmers' ratings of constraints to using organic agriculture*

the IMA's regional head for the Zona da Mata stated that it could help farmers with certification, including soil testing. Again, some farmers might have been aware of this and others might not. In the case of labour, it is not clear that organics would demand greater labour input, but there was such a wide spread of responses that some farmers clearly thought they would.

The constraint 'No market for products' got a very low rating, and would have given an even lower one had one or two high ratings not affected the mean. Farmers clearly think there is a market for organic products.

Is this optimistic? The EMATER agent who was sceptical suggested that the market would be fine as long as only a few people were in the business, but would quickly become overcrowded. Besides, it is not clear that consumers would pay extra for organic milk, even if they would for vegetables; and even if they did, the low price of milk means that the difference would not amount to much. However, as reported at the beginning of this subsection, so far the market for organics has been excellent. In any case, farmers were motivated by sustainability concerns as much as the market; prices were not the only or even the main attraction of organic agriculture.

So far, organic agriculture looks like a good option in this survey. It may not always sequester carbon compared to conventional options, but probably would in the survey area, given the constraints to using chemical fertilizers. It delivers significant co-benefits in terms of reduced pollution from agrotoxins. It also fits with farmers' own perceptions and wishes.

However, farmers seemed to be basing their responses on the impression that they were already nearly organic because they did not use many chemicals. But they may not have thought about the drawbacks of maintaining soil fertility organically. So in practice more information might be obtained by comparing results for an individual practice associated with organic agriculture. The survey asked farmers about green manure outside the context of organic agriculture, but it is worth presenting the results together with organics in this subsection; although it could be adopted outside an organic framework, it would be very important as part of it. Also, although green manure would be mostly understood as mucuna by these farmers, they were asked about the potential for using another crop, pigeonpea, as a strategy for improving soil fertility. This is also presented here.

Although several farmers had experimented with mucuna, virtually nobody was using green manure on a regular basis; and although farmers were enthusiastic about green manure, and thought it good for the soil, fewer actually expressed a wish to adopt.

Why? The most significant constraint was lack of technical knowledge, but the difference between this and the next constraint down (displaces more profitable activities) was not significant. Perhaps more relevant is that only 57 per cent of the whole sample felt able to express an opinion on constraints to green manures; farmers simply did not know enough about their use to return consistent ratings, even though some had experience with mucuna.

There is one other constraint that is of interest, however, and that is labour. Its mean rating was not very striking, but there was a significant difference between the ratings allocated to labour by the Noroeste Fluminense farmers (7.5) and those in the Zona da Mata (4.7). These were small samples after disaggregation by site, but they quite closely match those given in the two locations for labour as a constraint to the farming system in general, which were 7.8 and 4.5 (see above). Moreover the ratings for 'Displaces more profitable activities' showed an inverse

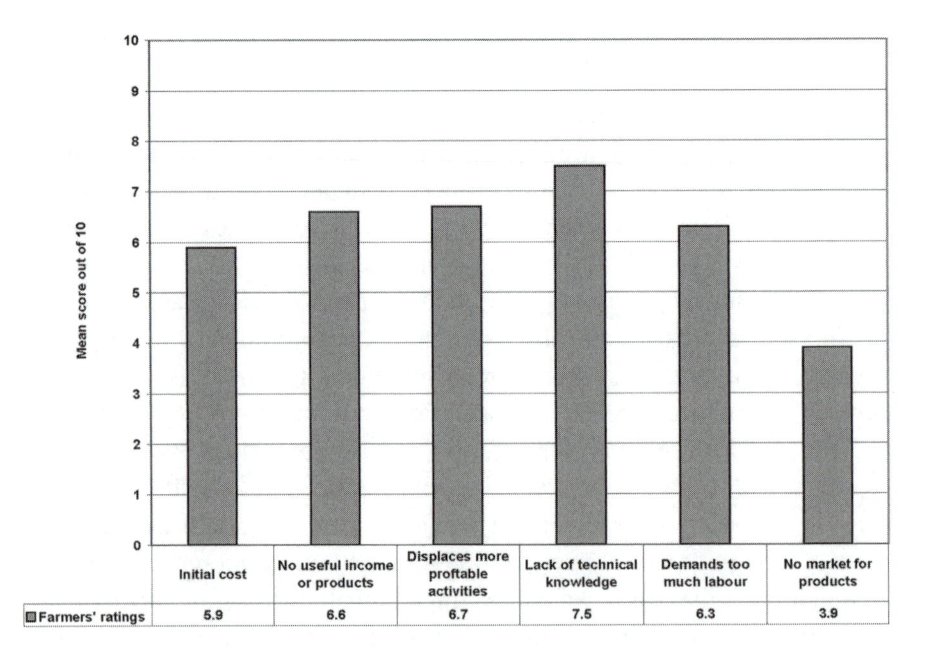

Figure 8.7 *Farmers' ratings of constraints to using green manure*

relationship to available labour, although again there was insufficient data for this to be significant.

In conclusion, lack of knowledge was the single constraint most mentioned, but there were other concerns, and the correlation of labour with its rating as a general constraint suggests it may be more significant than it appears. But in general, although the differences between some constraint ratings overall was significant, the inconsistent ratings suggest that farmers are unsure as to what the problems really are. It is therefore impossible to be certain that their enthusiasm for organic agriculture was based on a full appreciation of what is involved.

One specific possibility was discussed with farmers regarding green manure, and that was the use of pigeonpea. This is not a common crop in Brazil; it was originally domesticated in India, and nearly 85 per cent of production is still there (Giller, 2001, p143), but it spread to Africa about 2000 years ago, was then – like Brazil's pasture grasses – carried across the Atlantic with the slave trade, and is an important crop in the West Indies (ibid.). It is a versatile plant, as the leaves, stems, pods and grains are all fit for human consumption as well as providing fodder for livestock; stems are also used for fuel (Snapp et al, 1998, p192).

In Brazil, cultivation of pigeonpea on any scale is mostly confined to a corner of southern Brazil on the borders of Paraguay (Jones et al, 2004, p436). But its presence as a volunteer plant is quite widespread in the survey area, and some farmers commented that it was good in a salad, with chicken or with beans. However, they were not aware of pigeonpea's considerable potential as green manure. Giller (2001, p162) points out that when intercropped with maize, it continues to grow on residual soil moisture long after the cereal crop has been harvested.

However, the real potential for pigeonpea in the area might be for pasture improvement through its nitrogen-fixing qualities. There seems to be little experience of using pigeonpea in this way, certainly in the Americas. But De Alcântara et al (2000) report an experiment, on a dark-red Latosol on an experimental station in Lambari, Minas Gerais, on the potential for pigeonpea and sunn hemp (*Croatalaria juncea* L.) to rehabilitate degraded land planted to *Brachiaria*. Soil sampling at six depths 90, 120 and 150 days after cutting suggested that pigeonpea was effective at fixing and mobilizing nutrients after the shorter period, but that sunn hemp was more so after 120 days and that neither were effective after 150, the extra nutrients having been mineralized.

De Alcântara et al concluded that green manures were effective in the mobilization of nutrients for the following crop and in improving other soil physical properties, but only if the planting dates of both crops were chosen with care (De Alcântara et al, 2000, p285). They also reported liming before planting and treating the seeds with an inoculant specifically produced by EMBRAPA Agrobiologia (op. cit., 279). This suggests that, while a carbon sequestration project using green manure might get excellent results rehabilitating degraded pasture or cropland, it would need to be very well organized.

As to whether the farmers would adopt, most were reluctant to give an opinion; although pigeonpea was frequently found on their farms, they neither thought nor knew much about it. No one was willing to rate it on the basis of experience. But 30 were willing to allocate ratings for 'would like to use', giving a mean of 6.6 out of 10. They did think it would be good for the soil (7.7).

For constraints, only just under half the farmers were willing to give ratings; this was even fewer than with green manure. The rest insisted that they did not know enough about pigeonpea and had no opinion. The different reactions to the idea of pigeonpea as a useful crop can be seen in the highest-rated constraint, 'No useful income or products'. The mean was 7.5, suggesting that farmers were not clear as to what pigeonpea would be grown for. The second highest constraint (although the difference was not significant) was 'Displaces more profitable activities', with a mean of 6.9.

Asked directly about labour as a constraint, farmers gave a rating of 5.6; the next rating down was for initial cost. However, there was a wide spread of ratings for both. Some inconsistency would be explained by farmers not knowing what initial costs there would be. In fact, for pigeonpea to be really effective for rehabilitating degraded land, the organic material should be incorporated into the soil after cutting, and this would involve labour and machinery; it would also be necessary to keep cattle away from the crop. So those who envisaged high costs and labour input were probably right.

In general, however, the main constraint to adopting pigeonpea is negative; farmers do not see many big problems in using it – they just do not see why they would want to. They also seem unsure as to why they would be growing green manure, or how to do it. Yet organic techniques would be needed for maintaining soil fertility if organic agriculture were adopted. There is a warning here that farmers might show enthusiasm for organic agriculture that might not extend to all the practices its successful use might demand.

Moreover, there is the question of whether organic farming in the survey area would have a worthwhile net mitigation effect on climate change. There would be a reduction in emissions from nitrogen fertilizer, but many farmers were not using much anyway. As to carbon sequestration, this would only be significant with widespread use of green manure, including pigeonpea and/or *Croatalaria* for pasture improvement; this might indirectly build up SOC by increasing NPP in degraded *Brachiaria* pastures, but as discussed earlier this might need careful timing. Last but not least, a switch to organics might enhance the on-farm sink over the current baseline, but perhaps not compared to the net sequestration possible through careful use of inorganic fertilizer. For climate mitigation, well-targeted use of nitrogen fertilizer for pasture improvement looks a simpler, more effective and less risky option.

But conversion to organic farming could be part of a wider PES scheme in which carbon sequestration was just one of the environmental goals, along with watershed management, biodiversity conservation and reduction of agrotoxins in the water supply. In that context, it *would* make sense in the survey area, especially if accompanied by pasture improvement using green manure, forage legumes and perhaps leguminous trees (discussed below). A programme of this type would be especially appropriate in the Zona da Mata, where farmers were clearly interested in organics, and could perhaps increase use of green-manure intercrops in maize, raising the output from the tomato/maize rotations that they already use. However, such a programme should probably be preceded by on-farm trials to ensure that these components fit into the local farming system – especially as regards labour inputs – and that an adequate certification system is in place.

Leguminous tree species

Trees within a system can be highly beneficial. According to Giller (2001, pp237–238), benefits can include mitigation of soil erosion through the provision of soil cover, mulch from tree prunings, and reducing nutrient losses by extracting nutrients from the soil and returning them to it as leaf litter.

The use of leguminous trees to sequester carbon and rehabilitate degraded pasture is suggested by two experiments described by Boddey et al

(2006, pp314–315), one at EMBRAPA Agrobiologia near Rio de Janeiro, the other at Angra dos Reis, an hour or so's drive down the coast. The first concerned a slope where the top 40cm of soil had been removed for the construction of a dam and was rehabilitated using *Mimosa caesalpiniifolia*, *Acacia auriculiformis* and *Pseudeosamanea guachapele*. The result was an increase in SOC from 44.5Mg/ha C to 65.7Mg/ha C under the *Mimosa* and 99.5 and 94.6Mg/ha C under *Acacia auriculiformis* and *Pseudeosamanea guachapele*, respectively. So the three tree types had restored the soil C stocks by 21–55Mg/ha C over a 15-year period. Satisfactory results were also recorded at Angra dos Reis, where the topsoil had been removed from a 50 per cent slope to provide foundations for a shopping centre; here there was an improvement of 2.1Mg/ha C a year. Ferrari and Wall (2004, p72) explain that they are also good for rehabilitating land that has been heavily compacted and that this has been proved in areas degraded by mining operations, which can be affected by compaction caused by heavy machinery. This is also a feature of soils that have been ploughed (ibid.). As stated earlier, the pasture soils of the Atlantic Forest biome are often badly compacted and that is one of the reasons for using *Brachiaria*, which can establish itself in such soil.

Leguminous, or other nitrogen-fixing, trees could not only assist rehabilitation of pasture but have other advantages, possibly providing forage but also shade for cattle – an important advantage in deforested areas where animals can be very exposed. Primavesi and Primavesi (2002, p12) state that, in the colder season, it is 3 to 4°C warmer under trees and in the hot season it is 3 to 4°C cooler. 'Pastures with at least 50 shade trees/ha,' they say, 'allow a yield increase of 15 to 30 per cent milk and around 20 per cent meat.' Farmers are aware of this need; two in Noroeste Fluminense specifically mentioned this as the main attraction of the trees. Both were in the Muriaé valley, where there seemed to be very few trees.

The trees could also provide fence posts. Evans (2004, p35) found that cattle-farmers in São Paulo State had a problem with cattle entering surviving areas of forest, where they not only damaged the vegetation but also got stuck and lost and occasionally even died. Yet the farmers found fencing was sometimes too expensive: Rs540, about £132, just for the wire for 600 metres, at the time more than twice a farm labourer's monthly wage. In light of this, the provision of materials to fence forest edges could provide both social and conservation benefits. A further benefit would be firewood; although most of the farmers in Itaperuna appeared to use bottled gas, those in Minas Gerais often used a traditional *fogo da lenha* on the back verandah, which also provides hot water by convection.

It would be easy to be too optimistic about these trees, as they have yet to be tried on a large scale in the region. The high SOC sequestration cited for the rehabilitation experiments was on soil that was very low on carbon to begin with, as Boddey et al (2006, p334) point out. And trees that are meant to fix nitrogen may depend on the presence of compatible rhizobia or, in their absence, inoculation; farmers should not be expected to assess this for themselves. As to their effect on soil fertility, Giller (2001, p233) points out that there is 'a dearth of information on the amounts of nitrogen fixed in the field by trees, mainly due to the enormous task of destructively harvesting and then analysing mature trees to estimate their total N content'. It may also be that the benefits are indirect; there is 'little evidence of substantial transfer of N from roots of legume trees to companion grasses, unless the tree is cut back' (Giller, 2001, p238).

Box 8.4 A place to try the trees?

In the Muriaé valley in Noroeste Fluminense, one older farmer was staying on a 23ha farm that he had not been able to run himself for several years due to injury; it was worked by a single employee. The farmer had improved some of this with *Brachiaria* but it was mostly elephant or Napier grass (*Pennisetum purpureum*) and other grasses referred to locally as 'native', although they are often of African origin. He also had about 0.3ha of maize and 0.5ha of sugar cane, the latter fertilized with urea to provide feed for dry periods. He didn't think farming was a good living and said that out of Rs2000 (about £524) return a month, he'd be lucky to get back Rs70 (£18) a day for himself, and claimed that it cost about Rs40 a day to have someone skilled working on his house. Ordinary labour cost less, about Rs12–15 a day, but he said it was often not available as many people had gone to the city.

It may be that this farm, with its lack of diversity, use of 'native' pasture species, lack of trees for shade and apparent lack of funds for investment, would be a good place to hold on-farm trials of practices such as pasture improvement with pigeonpea and planting leguminous trees for pasture improvement and shade. This might also permit on-farm analysis of their contribution to nitrogen fixation and soil structure in the local pastures.

Boddey et al (2006, p334) are cautious in their assessment of how much SOC could be accumulated when rehabilitating degraded pasture in the real world. However, they suggest that if degraded pastures typically have an SOC content to 60cm of about 80Mg/ha C, then leguminous trees might increase this by about 25Mg C over a 20-year period. This does not represent a huge windfall in terms of carbon prices on soil carbon alone, but the tree biomass itself would also represent carbon storage, at least on a temporary basis. Moreover, the land would be serving other purposes during this time. So would the trees. If any carbon payment is seen as a small but useful subsidy for a land use that is profitable for other reasons, then it may be worthwhile.

This is suggested by the survey results, which revealed a cautious enthusiasm (it should be remembered that the farmers were not told carbon payments would accompany any of these practices). Only three farmers were using the practice already, and all three cited prevention of soil erosion as a reason for adoption. However, the remainder of the farmers knew little about the use of leguminous trees.

They were told that the trees grew in 8–10 years, gave shade for cattle, were good for soil fertility and were useful as fence posts. On this basis, the level of interest in adoption was high. About one farmer in four gave the maximum rating of 10, although one in three expressed no opinion. The farmers also seemed to think that the trees would be good for the soil (8.2 out of 10), although again over a third of respondents gave no opinion.

Given the high level of interest, and the enthusiasm of some EMBRAPA researchers, it seems especially important to know what constraints were perceived by the farmers. The top-rated constraint was 'Lack of technical knowledge' at 7.5. This was higher than for most other practices.

But the second, 'Displaces more profitable activities', was not far behind at 6.6. Those who marked this high were probably thinking of space rather than labour, as labour demands of leguminous trees would not be very high. One farmer with an 8-acre farm specifically mentioned space. Initial cost (6.6) might be more connected with labour. But it might also reflect uncertainty as to how much seedlings might cost.

There is also a constraint that farmers would not have known of, and that is assessing whether the trees did fix nitrogen and whether inoculation with rhizobia was required for them to do so. As the farmers cannot be expected to analyse soil biota, the practice might demand some input from

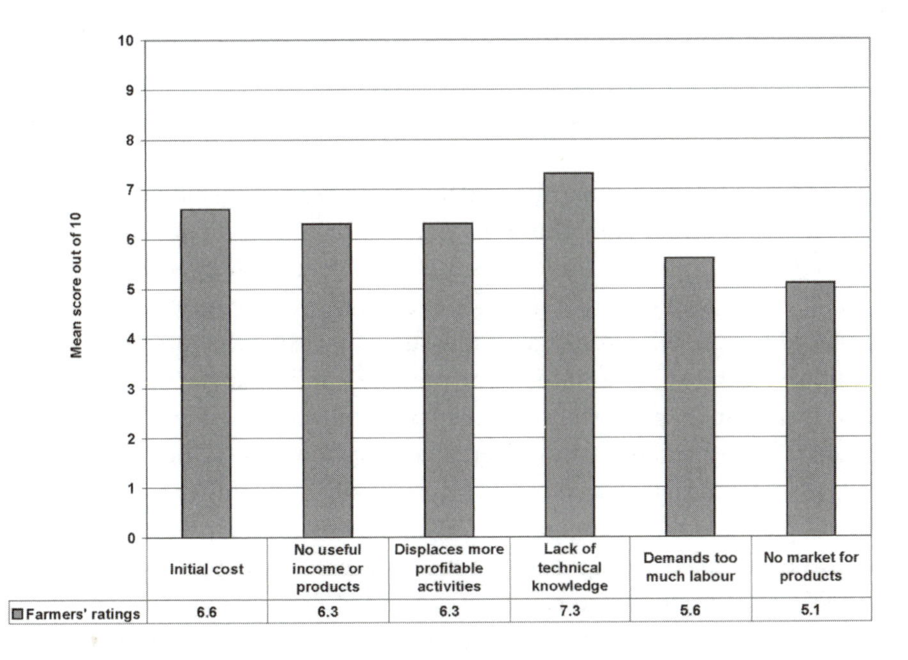

Figure 8.8 *Farmers' ratings of constraints to using leguminous trees*

extension services to attain its full potential. Another problem might be seed supply, raised by two farmers in the Zona da Mata.

Farmers liked leguminous trees on the basis of what they were told, but, as one of them said: 'I have an open mind; I just don't know what the pitfalls are.' Leguminous trees could sequester a useful amount of carbon both in their own biomass and through pasture improvement. Even so, to make sure the leguminous trees do fit with the farming system, on-farm research in any project area would be a wise start.

Conservation tillage and rotations

This subsection looks at two management practices that improve soil structure and reduce erosion through agronomic practices, and thus increase the size of the carbon sink. The practices are conservation tillage (CT), involving the reduction or elimination of mechanical soil preparation, and sustainable rotations – that is, cropping sequences that do not mine nutrients and organic matter. They are being presented together here because they cannot easily be separated in the Brazilian context, for reasons that will become clear. In fact, despite the widespread use of zero tillage or CT elsewhere in Brazil, it has not hitherto been much used in the survey areas; it would be applicable there, but probably not in quite the same way. This is a slightly complicated situation, and some discussion is needed.

Reduced tillage regimes go under several names, including CT, no-till (NT), conservation agriculture (CA) and direct drill; the last is normally used in Brazil (*plantio direto*). These names do not always mean the same thing. Conservation agriculture, for example, strongly implies not simply reduced tillage, but a package of measures including the retention of crop residues. In fact CA has three basic premises: minimum or no mechanical soil disturbance; permanent organic soil cover, being either the crop or a mulch of crop residues; and diversified crop rotations (Giller et al, 2009, p24). Such packages can be relatively inflexible, making it hard for farmers to pick and adapt components as it suits them; the problems this can cause have been highlighted in a recent landmark paper by Giller et al (2009), and are discussed briefly at the end of this section.

However, the term CT, for conservation tillage, has been used here as it does cover most reduced-tillage regimes, including that used by the IPCC. Indeed CT is recognized by the IPCC, whose definition of it is that of the Conservation Tillage Information Center (CTIC). This calls for crop residues to remain on the soil surface after planting; the amount or percentage of residues that must be used in this way depends on whether water or wind erosion is seen as the primary threat. Under the CTIC definition, certain forms of tillage are acceptable if they are compatible with these requirements for crop residues. Other bodies call for different procedures that eliminate tillage altogether, and this would generally be required by what is called conservation agriculture (CA) as laid down by the European Conservation Agriculture Federation (ECAF). However, any reduction in inappropriate tillage should reduce organic matter breakdown, and thus slow loss of soil C even if it does not sequester it. Since this also increases productivity in the long term, funding conversion to no-till through carbon credits should be a win-win situation – especially as adoption may sometimes be constrained by lack of information about the management practices required, as Díaz-Zorita et al (2002, p15) found in Argentina. In countries with weak extension services, carbon-related funding could fill the gap.

Farmers have practised tillage since ancient times. Its main purposes are to prepare a seedbed and to control pests and diseases by exposing them to the air. It also exposes SOM to the atmosphere, stimulating mineralization which releases nutrients to the crop – but also causes the loss of carbon as CO_2. It is this as much as soil erosion that has brought about agriculture's historic contribution to the atmospheric carbon pool. But most tropical soils, it has been argued, need not be tilled (FAO, 1998, p1). An easier strategy is to leave crop residues in the field, where they will encourage a build-up of soil biota that will – it is claimed – aerate the soil just as well (FAO, 2002a, p1).

If this is the case, then the supply of nutrients to the crop will be increased, not decreased, as the crop residues will be incorporated into the soil as SOM/SOC by increased biotic activity. For this to occur under CT or NT, there is a need to understand how they affect nutrient cycling; the process of mineralization must not actually stop, or the farmer will simply be left with immobilized nutrients in the form of crop residues in the soil.

Mrabet et al (2001, p506) points out that in many dryland forage-based cropping systems, more than half the nitrogen required by the crop comes from mineralization of SOM. As Albrecht (1938, p1) expresses it: 'Attempting to hoard as much organic matter as possible in the soil, like a miser hoarding gold, is not the correct answer. Organic matter functions mainly as it is decayed and destroyed. Its value lies in its dynamic nature.' This applies especially where there is potential for immobilization of nitrogen in low-quality residues built up in nutrient-poor soils, and this appears to be the case in the pastures in much of the survey area, with high C:N ratios due to the build-up of nitrogen-deficient root material. So the objective, with CT, is to build up the percentage of organic matter in the soil rather than stop its mineralization completely, and in the short term extra inputs may be needed.

Provided the organic matter is properly broken down, however, there should be improvements in soil structure; as Lal (1997, p94) explains, microaggregates develop around decomposing organic matter. As discussed in Chapter 2, higher SOM/SOC content is associated with soil that is more aggregated, as the aggregates protect the organic matter from mineralization. They also increase infiltration capacity relative to more compacted soils; this promotes greater water availability in the root zone. As higher SOC content also increases resistance to compaction, there is a circular effect, with higher biomass, greater SOM and greater resistance to compaction – a phenomenon assisted by the reduced passage of heavy farm machinery.

The resulting better-aggregated soils are less prone to water erosion due to their higher infiltration capacity; this also reduces loss of SOC. CT is therefore seen as an important strategy for enhancing the carbon sink. As Lal and Kimble (1997) state: 'Conservation tillage is known to enhance SOC in the surface soil horizons through several mechanisms (e.g., alterations of soil temperature and moisture regimes, and erosion control)' (Lal and Kimble, 1997, p245).

However, CT's origins had nothing to do with carbon sequestration; it was a response to declining productivity due to the impact of mechanization (FAO, 2001b, p1; Sisti et al, 2004, pp39–40). It began in the United States, but spread quickly in parts of South America in the 1970s and 1980s. Jose Benites of FAO has recorded that, for many farmers in the southern Brazilian State of Santa Catarina, it was 'a stark choice between soil conservation and starvation' (FAO, 2001b, p1). But as mentioned in a previous chapter, farmers in the southern states of Brazil are increasingly aware of its potential for climate mitigation.

Reduced tillage regimes should not be seen as a miracle solution. They do in general favour the build-up of soil C, but it is not difficult to find examples in the literature where this has not been the case. In fact, whether or not CT does build up C seems to depend on rotations and their effect on the overall nitrogen balance, root depth, and whether or not C is moved between the soil horizons. Less directly, although biomass production – and thus soil C content and productivity – can be enhanced, this will be subject to a series of interlocking factors that can change with time and management practice. Sisti et al (2004) compared soil carbon stocks under zero and conventional tillage in a 13-year experiment in southern Brazil. Under a wheat/soybean rotation, there was no significant difference. Where a nitrogen-fixing legume vetch was included in the rotation, there was. Nitrogen was exported when soybean was harvested, but rotation with vetch appeared to compensate for this. However, when carbon stocks were higher, the gain appeared to come from root residues; this emphasized 'the importance of studies of root biomass and turnover to further an understanding of the driving forces behind carbon accumulation/sequestration under [zero-tillage] systems' (Sisti et al, 2004, p56). This seemed to confirm other studies that suggested CT did build up more SOC than conventional tillage when a green-manure legume was included in the rotation (Sisti et al, 2004, p50).

There is a further complication: reducing or eliminating tillage may increase SOC compared to conventional tillage in the surface layer but not the entire soil profile. This is because a conventional mouldboard plough will move soil carbon into the lower soil horizons. Although untilled soils should be less vulnerable to erosion, a higher percentage of their soil carbon will be in the upper horizons, and they will lose more of it to any erosive processes that do still occur. Lal (1997, pp97–98) goes so far as to suggest that ploughing of no-till plots might be necessary every five to seven years in order to sequester the SOC from the upper horizons into the subsoil, but adds that long-term experiments are needed on different soils, and in different ecoregions, to test this.

Against this, however, there is a potential benefit from reduced mineralization through reduced erosion under CT. It might be spectacular. Boddey et al (2003, p606) report that: 'In a recent review of the literature ... [it was] estimated that where ZT [zero tillage] had been introduced in Brazil there was a mean reduction of 75 per cent in soil loss and a 25 per cent lowering in water run off'. As stated in Chapters 2 and 6, there is enormous uncertainty about what percentage of the organic matter translocated is mineralized, but Lal (2003, p437) suggests it could amount to as much as 1.2Pg/yr C.

This link between erosion, CT and the carbon sink does not seem to be prominent in the literature, which tends instead to emphasize the build-up of SOM under CT through other mechanisms. However, the erosion factor suggests that CT's significance for climate mitigation could be considerable even without the build-up of SOC expected through improvements in soil structure and the reduction of mineralization during tillage. This has been demonstrated by Doraswaimy et al (2007) in Mali. They suggested that 20kg/ha was being lost under conventional land preparation, and that conversion to ridge tillage would eliminate this loss (Doraswaimy et al, 2007, p70). The amount of loss prevented would obviously be completely different in other ecosystems – Mali is a very different proposition from the Atlantic Forest – and reduced erosion does not constitute carbon sequestration. But it does reduce CO_2 emissions relative to business-as-usual.

In sum, it should be accepted in principle that CT is an effective strategy for enhancing the agricultural sink, provided that the caveats above on nutrient cycling, depth of SOC and the need for rotations are borne in mind when any project is designed.

How applicable is CT to the survey area? Although most experience with CT is with arable farming, it can also be used to establish *Brachiaria* pastures. But there is mixed evidence as to how helpful this would be. Sotomayor-Ramírez et al (2006, p689) suggest that adopting CT in this context could be good for soil quality, decreasing SOC loss, and could lower costs. However, use of NT, minimal and conventional tillage in establishing *Brachiaria decumbens* in an experiment in Puerto Rico did not indicate any short-term change in the soil's capacity to lose carbon and nitrogen. Sanabria et al (2006, p433) compared three tillage regimes in the restoration of a degraded pasture of *Brachiaria humidicola* on a sandy loam soil at Monagas in Venezuela, and found that using a conventional harrow gave the fastest dry-matter yield, although the other two treatments – NT, and ploughing plus subsoiling – were equal in the end. However, these results might or might not apply with a lower level of inputs (they included two annual applications of urea at 100kg/ha). The authors quote studies (Sanabria et al, 2006, p419) suggesting that changes in soil physical properties, including compaction due to excessive ground preparation, are a major cause of pasture degradation, but CT or NT may not be the right answer for this if there is insufficient release of nutrients and consequent organic matter build-up.

Besides, tillage is practised in part for weed control – as is fire; Evangelista et al (2001, p17) point out that burning is often used to control *macega*, or cordgrass (*Spartina densiflora*) but can have drastic consequences, increasing soil erosion and reducing palatable species as well as weeds. A farmer showed the author one patch in the Zona da Mata where a neighbour had used this tactic, but the pasture had not recovered afterwards, and had now been abandoned to woody encroachment – which was perhaps not bad for carbon stocks, but of little use to the farmer. It may be that at least some ground preparation is a better idea, and Carmona and Zatz (1998, pp1515, 1523) found in an experiment on establishment of *Brachiaria decumbens* elsewhere in Minas Gerais that at least one deep operation was the most economic option for controlling weeds; in this case, it was also found that zero tillage was less efficient in pasture establishment.

In any case, the farmers surveyed were not using CT much and, although the few who were rated it highly (9.2 out of 10), the question 'would like to use/use more' scored much lower at 5.5 – the lowest mean for this question for any practice in this survey. Besides, of the six who were using CT, three were neighbours with smallish farms and two of them were less dependent on pasture than was typical for the region, growing fruit and vegetables as well, while the third stated that he did not improve the pasture because he was a tenant and would derive no long-term benefit. So none of them had much reason to plough anyway.

However, the other three were pasture farmers who were more typical of the region. One said that he used to use minimum tillage but had never heard it referred to as such. Another 'ploughed lightly' because some of his land was sloping, and gave erosion as his reason for using minimum tillage. The third, his neighbour, also said that he used minimum tillage to avoid erosion, which he thought affected about 30 per cent of his land. He was in the process of planting with Angola grass (*Brachiaria mutica*) and was expecting to sacrifice about 20 per cent of yield by using minimum tillage, so clearly regarded erosion as a serious problem.

The implication might be that a few farmers had specific reasons for not tilling much, either because they didn't need to or because their land was sloping – and that without such a specific reason, people preferred to plough. But although 'Would like to use/use more' had a low mean, there was little consistency. Indeed many farmers preferred not to give a rating at all because they had not heard of the practice before. This suggests caution in scaling up this finding.

From the foregoing, it would seem that there is not much point in pursuing NT or CT as a sequestration option in the survey area, despite its importance elsewhere in Brazil. The farmers are not especially interested unless they perceive a big problem with erosion, and anyway, it is not clear that it is really helpful for pasture establishment. However, there are three reasons why further research into the applicability of CT in the survey area might be worthwhile.

First, pasture does tend to be on sloping land, with the flatter valley bottoms cultivated. Given the frequent signs of erosion observed, it may be wise to encourage very cautious ground preparation during pasture establishment (or, better still, contour planting; but there are some problems with this – it is reviewed next).

Second, minimum tillage is perfectly practicable for tomatoes. Only one farmer in the Noroeste Fluminense was growing them, but in the Zona da Mata most were – usually on a small scale, but one had 4ha, and a farmer in the neighbourhood who was not interviewed was quite a large commercial producer. EMBRAPA (2003) describes a system of direct-drill through cereal straw;[3] this can be grown specifically for the purpose, in which case EMBRAPA suggests it be planted 55 days before transplanting. However, straightforward rotation can also be used, typically with rice – a few of the farmers surveyed grew this, or had grown it in the past – or maize, which was quite often rotated with tomatoes and sometimes used for fodder. The residues are then ploughed in. Farmers in the Zona da Mata are well aware of the value of this practice for soil fertility, and one commented that if he used the residues for silage instead, he really noticed the difference.

It is not clear whether the farmers could be persuaded to leave the residues on the surface instead and direct-drill through them. There is a need

Box 8.5 Small-farm sustainability

In the Zona da Mata, cultivation was mostly in valley bottoms, with sloping land used for pasture. This clearly makes sense, as ground preparation on steep slopes could cause erosion, and the valley bottoms would have more moisture and nutrients. However, one farmer in the Zona da Mata had a small area – just 2ha, and there were four people working on the farm, so he had to use this land intensively to support them. He was one of the few to admit to commercial cultivation on sloping areas.

But much of the farm was pasture. He thought farmers would not use CT for pasture establishment; they did have to plough, he said, to establish *Brachiaria* in native pasture. On slopes, he did do pasture improvement by hand, but this was very expensive and would have been because of the danger of using a tractor on slopes as much as to conserve the soil. However, being forced to cultivate sloping land, he was very aware of soil-conservation issues. He rated erosion higher as an overall constraint than most farmers, giving it 9 out of 10, and practised contour planting, according it a rating of 9 for soil health.

Like many local farmers, he also planted maize after tomato so that it would get residual benefit from the fertilizer. He reported that EMATER had given 65 local farmers some fertilizer this year, but this was unprecedented and he didn't know if it would happen again. He would use his on maize, suggesting a perceived need for fertilizer on this crop although it was normally reserved for higher-value crops such as tomato. But, like a number of his neighbours, he was concerned about the use of chemicals and knew the public were too. He was interested in the idea of pigeonpea as a green manure. But he thought space for it would be a problem on his small farm, and this would presumably apply to other green manures and to leguminous trees as well (he rated them at 8 and 9, respectively, for 'Displaces more profitable activities'). He was an interesting contrast with the Noroeste Fluminense farmer described in Box 8.4, who had enough land but no one to work it, nothing to invest in it and no obvious wish to improve sustainability. This farmer, by contrast, had enough labour and a keen awareness of sustainability issues but little space for the management practices that might address them.

to control weeds, but the rotation and the straw should protect from this, and in any case the Zona da Mata farmers seemed more concerned with pests and diseases than weeds – one mentioned having to use two types of fungicide. One farmer expressed doubts about direct-drill because of the extent of soil compaction, but the EMBRAPA method does include fertilization in conjunction with what is effectively a small subsoiler working at 20cm. Another farmer thought direct-drill might save money as they currently have to break up the maize stalks and this already requires machinery, which is costly.

In fact, an experiment in direct-drill of tomato in the southern State of Santa Catarina, reported in the magazine *Agropecuária Catarinense*, there was a reduction of 60 per cent in fertilizer costs and a reduction in labour; one family reported production costs of Rs1.87 (about 49p) per box against Rs6 to 7 (£1.56 to £1.82) in conventional production (Tagliari, 2003, pp27, 29). This experiment took place in a region where there was widespread monocropping of horticultural crops, mainly tomato, with heavy chemical inputs; the farmers in the Zona da Mata, by contrast, already have more sustainable practices for tomato, and there may be less to be gained. On the other hand, this also means that direct-drill would be a less alien concept for them than it was for the Santa Catarina group, as the Zona da Mata farmers already use cereal straw to maintain fertility. So a small-scale experiment in the area might be worthwhile, although it would not have a huge impact on carbon stocks.

However, the third reason why minimum tillage might be important for the area is that it is being advocated, with appropriate rotations of cereals and legumes, as a good way to rehabilitate pasture. In carbon terms, this would not be relevant where the pasture is in good condition. As mentioned earlier, well-managed pasture can maintain or even augment SOC content of the original forest cover. Where the pasture is *not* in good condition, however, SOC will be very inferior compared with other land uses. Valpassos et al (2001, pp1539–1545) compared a 20-year pasture, unfertilized since establishment, with native cerrado, zero tillage under a rotation of beans, oats, soya and maize, and conventional-tillage rotation of soybean and corn using incorporated crop residues. The pasture had an SOC content of 22.86g/dm^3; even the conventional tillage was higher at 24.15g/dm^3, while the native cerrado had 30.57g/dm^3. However, the no-till rotation came top at 42.52g/dm^3 and had a markedly higher microbial biomass than any of the other land uses and a lower bulk density than all but the native cerrado. Moreover, there is no reason to suppose that it had necessarily reached equilibrium – that is to say, it might yet accumulate more SOC, whereas the degraded *Brachiaria* pasture would not be likely to.

So the best way to build up carbon stocks – and perhaps profitability – may be to introduce rotations with crops, planted with CT/direct-drill so that extensive soil preparation does not worsen already serious soil compaction. In a paper on recovery of degraded pasture published by the Universidade Federal de Lavras (UFLA) in Minas Gerais, Evangelista et al (2001, p15) advocate this solution: 'With the advent of direct-drill,' they say, 'the technique of cultivating degraded areas of pasture becomes more attractive every day.'

This optimism should perhaps be tempered, for a number of reasons. First of all, Valpassos et al's figure is only for the top 10cm of the soil profile; as discussed earlier, CT may be less effective compared to other management practices at accumulating SOC in lower profiles (Baker et al, 2007). For the farmer, this may not always be important, but it could affect the permanence of the sink. However, once again, caution is needed. There is now some evidence that CT may actually accumulate *more* carbon in the lower horizons, especially where rotations include legumes with deep root systems. Recent work by EMBRAPA at five sites in the southern State of Rio Grande do Sul has found that where no-till had been in operation for 14–18 years, soil carbon stocks under zero till (ZT) were 8.8 to 16.9Mg/ha greater than under conventional tillage; moreover, 32 to 92 per cent of this was below 30cm, leading the researchers to suggest that sampling to only 30cm could underestimate carbon sequestration under zero tillage in tropical soils by a factor of between 1.5 and 4 (Boddey, 2007, pers. comm.). It does seem that CT is a classic case of more research needed on a topic that, before climate change, was not seen as important.

Moreover, as discussed earlier, the effect of tillage regimes on CT may depend on the rotation used. There is no point in using CT to rehabilitate pasture if the rotations used are not sustainable. Those cited by Valpassos et al (2001) used a cropping sequence that would accumulate some SOC through root biomass and might fix at least some nitrogen, but continuous cereals would not.

It is also important to investigate farmers' attitudes to rotations to establish why they did not seem to find alternative land uses for relatively unprofitable, often degraded, pasture. Farmers were asked about rotations separately to CT. However, if CT were used to rehabilitate pastures, then sustainable rotations would be essential, so the two topics should be considered together, rather as organic agriculture and green manure were earlier in this chapter.

First, what were the perceived constraints to minimum tillage? The highest-ranked constraint was lack of technical knowledge at 6.6, but the difference between this and the other constraints was mostly not significant. Moreover responses were very inconsistent, reflecting the fact that some farmers had heard of the practice but some had not.

This is surprising when about half of Brazil's cereal area is now worked this way (Esteves, 2007, p1). Moreover one of the few farmers who did know about mimimum tillage thought it was for large farmers only. This is less surprising, as according to the Brazilian direct-drill association FEBRAPDP,[4] in 2005 small farmers were using it on just 600,000ha while medium and large farmers had applied it to 24.9 million hectares (Esteves, 2007, p1). It is only now spreading to smaller farms, where it can be applied with animal traction; according to Esteves, this means there is no need for costly machinery hire. This would be attractive to the farmers in the survey, who faced bills of up to Rs85 (about £22.11) an hour for

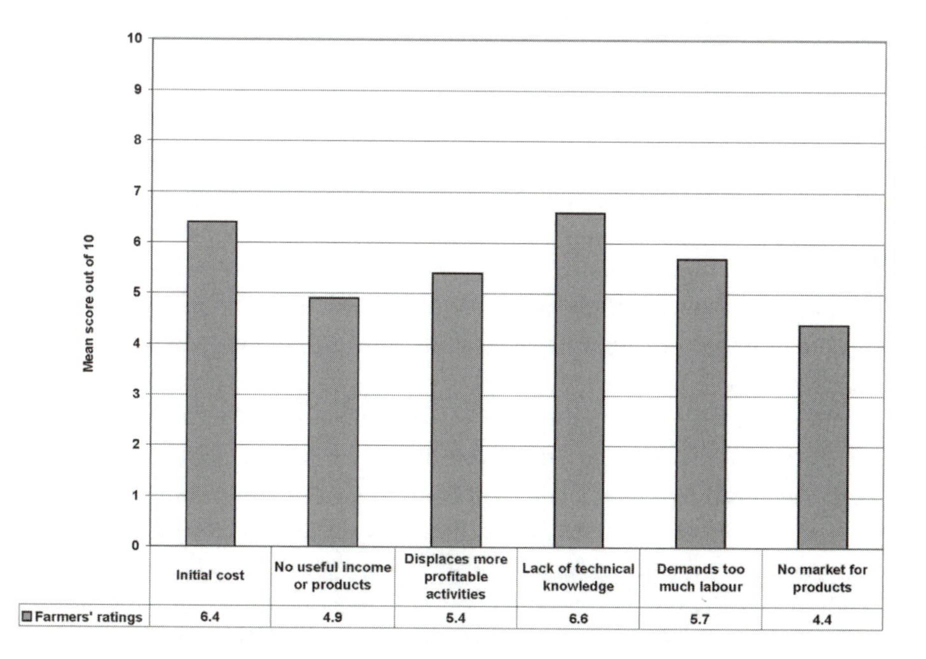

Figure 8.9 *Farmers' ratings of constraints to conservation tillage (CT)*

tractor hire. Despite this, they put initial cost as the second highest constraint, albeit with a wide spread of ratings. But in the main, the means for constraints to minimum tillage do not convey much because the farmers were so unsure, and had such a wide spread of views, that their ratings should not be scaled up.

However, as discussed, CT/direct drill would be more relevant for many of these farmers if used for rotating cultivation of degraded pasture. If this were attractive to them, they could be introduced to CT as part of a technical assistance package.

Moreover, whereas the farmers had little to say about CT during the interviews, making it hard to unpack what the ratings really meant, this was not the case with sustainable rotations. Just over half the farmers answered the question on how much they liked using rotations. They rated the practice quite high at 8.8 out of 10. Nearly as many stated how much they would like to use them/use them more (7.3), but there was wide variation in the ratings, with farmers either wanting to use the practice and rating it high, or in a minority of cases ranking it very low. But there was general agreement that rotations were good for the soil. So half the farmers used rotations, half did not and those who did not were divided over whether they wanted to, despite giving it a high rating overall. This suggests one or more serious constraints.

However, the ratings for constraints in Figure 8.10 do not suggest this, with the constraints generally marked low; in fact, every single rating was below the mean for that constraint across all the practices. 'No market for products' is rated very low, so farmers either think they can sell whatever would be grown, or assume it would not be for sale (for example green manure). The highest constraint is 'Displaces more profitable activities', presumably pasture. This is significantly greater than 'No market', but not significantly greater than the next constraint up, which was initial cost. Moreover, there was a wide spread of ratings for all constraints.

But it is interesting that the widest variation of all was on the rating for labour. While nine farmers rated this at >8, the remainder of ratings were clustered between 2 and 4. Disaggregation suggests that farmers in Noroeste Fluminense regarded this as a greater constraint. Also, the highest-rated constraint for Noroeste Fluminense was 'Displaces more profitable activities' at 6.5, against 3 (SD = 2) in the Zona da Mata. As their farms were bigger than those in the Zona da Mata, this reflects demand for labour.

Farmers' comments bear this out. One or two remarked that they had grown crops in earlier years, but no longer could (for example rice). One farmer mentioned the labour shortage specifically in the context of the rotations question. He had been farming for 44 years (since he was six years old, he said) and remarked upon the loss of labour over the last 40 years.

Two farmers in Noroeste Fluminense mentioned water-related constraints; two in the Zona da Mata also raised it (they were neighbours; one was replanting trees to deal with the shortage). A third said he thought it was a problem for other farmers. It didn't seem to be widespread in either location, but it would clearly be unwise to plan any drive for rotations without ensuring that there was enough.

It would also be easy to conclude that, while rotations were not an easy carbon-friendly tactic for Noroeste Fluminense, they were for the Zona da Mata, where they were already popular. However, although farmers there were happy using rotations, few evinced any great enthusiasm for using them *more*. This might be because, while their extra labour allowed them to work the land more intensively, they were already doing so as much as possible, and there would not be the capacity to rehabilitate degraded pasture in this way. If the rotations are not used for this, there is little carbon-related rationale for encouraging them instead of simple pasture improvement.

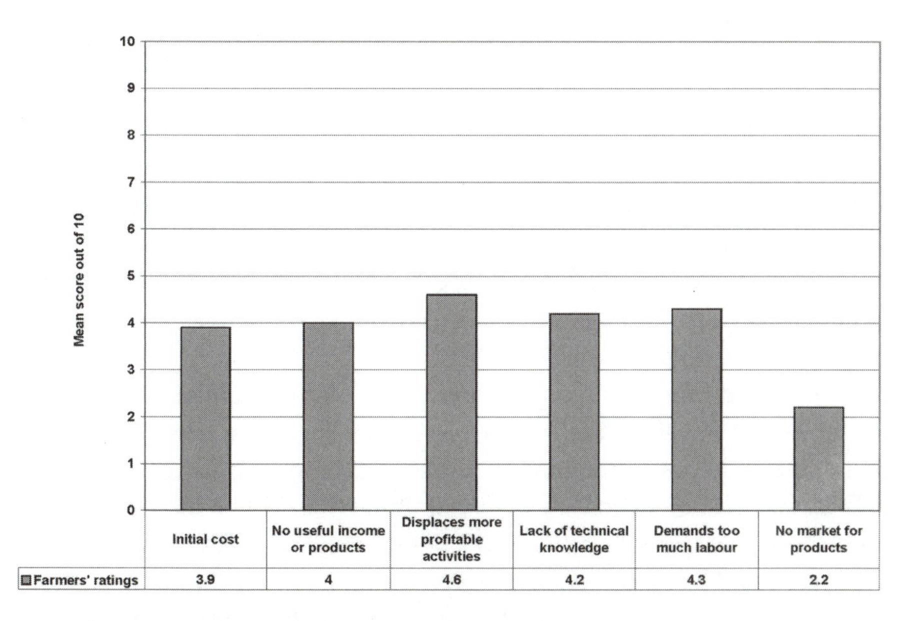

Figure 8.10 *Farmers' ratings of constraints to sustainable rotations*

CT has received much attention in Brazil as a potential climate mitigation strategy under the CDM. But it looks a weak option for the survey area except in certain specific applications (it might be good for tomato, for instance). There is a question mark over water for the rotations that would be needed, and the labour is not really available.

Some of these problems should not be surprising in the light of an important 2009 paper by Giller et al, who examined the agronomic practices implied by reduced tillage regimes with reference to their potential for smallholders in Africa (they looked at these regimes under the banner of conservation agriculture, or CA, rather than CT). Their conclusions, if not entirely negative, are somewhat bleak. Points raised include the fact that in sub-Saharan Africa, '[U]nless there is a ready market for the grain, farmers tend to grow legumes on only a small proportion of their land, and certainly not sufficient to provide a rotation across the farm' (Giller et al, 2009, p28). As discussed earlier, Sisti et al (2004) found a complex relationship between carbon sequestration and nitrogen-fixing legumes under reduced tillage, and if farmers do not practise such rotations, they may not build up much soil carbon – other than that not emitted due to reduced erosion, a gain that Giller et al (2009, p29) do accept, though they point out that it will be long-term.

Giller et al (2009) have a number of other problems with CA. One is that the labour requirement will shift from herbicide application, normally seen as men's work, to weeding, seen as women's work – meaning that the labour requirement will be not simply reduced but re-gendered (op. cit., p27). The authors also argue that any build-up of organic matter in CA can be ascribed to the retention of crop residues, a practice that can be adopted without reduced tillage (ibid.). They also point out that retention or otherwise of crop residues is defined by a number of factors, and may not always be a profitable or practicable option (op. cit., pp25, 30). This point has already been discussed in this book, with reference to leakage in sequestration projects (see Chapter 5).

None of the foregoing means that using CT to establish rotations on degraded pasture should be rejected outright. The labour constraint is real, but one should not assume that it is impossible to *increase* the supply of labour at a given location – a big assumption that will be discussed in the next chapter. Regarding Giller et al's points, these are extremely important – but it should be noted that they were examining conservation agriculture as a technological package advocated by development organizations, and one of their objections to CA is that its proponents resist unbundling this package to see which components of the package are really important. In fact, a practical, well-run project would probably do exactly that.

In the author's opinion, packages involving one or more of the components of 'full' CA remain important options for degraded agricultural land, both pasture and arable. However, Giller et al are right to object to the way in which CA has been advocated as a 'magic bullet'. There is no such thing in agricultural development; and in the survey area, pasture improvement, leguminous trees and possibly organic agriculture are more obvious choices.

There are also other soil-conservation options, and two of these – vegetative strips, and contour planting – are reviewed next.

Managing the slope: Contour planting and vegetative strips

Ploughing and planting should always be done along the contour where possible. Up- and down-slope ploughing will speed up water flow, lowering infiltration and encouraging erosive processes. As has been discussed, this is a major source of carbon emissions from soil, albeit a difficult one to quantify.

Even if ground preparation does follow the contours, it must do so as closely as possible, or water will flow laterally along the furrows and will become concentrated at the lowest point, increasing its erosive capacity. This is as true of pasture improvement as it is of any other activity, and sloping areas are often used for pasture. Indeed, Fiori et al (2001, p75), in a study of erosive processes in the Pantanal Matogrossense on the borders of Paraguay, noticed that the worst erosion was in the most sloping areas of pasture.

The study area was no different in this respect, and the fact that pasture improvement was generally done at the onset of the wet season would not help. The EMATER agent in the Zona da Mata commented that the sugar cane previously planted on some of the hills *had* been planted on the contour; erosion had arisen since, because the pasture that succeeded it had not. However, he admitted that there was also some danger of overturning the tractor if ploughing along the contour. There appear to be few recent figures for accidents involving tractors, but Schlosser et al (2002, p977) surveyed a sample of tractor drivers in Rio Grande do Sul State and found that 39 per cent had had some kind of accident and that overturning of the tractor was responsible for 51.7 per cent of the cases involving serious injury. Extrapolating up for the whole of Brazil, it seems likely that quite a few farmers and agricultural workers are killed in this way, and reluctance to plant along the contour is understandable.

Moreover, Van Oost et al (2006, p457) state that contour planting will do little to improve the soil unless accompanied by other measures; they appear to be thinking of vegetative strips or barriers. As these need to be planted on the contour to be effective, the two practices are not easily separated, and they are presented together in this subsection. (The farmers were not asked about terracing, as the labour costs would not be justified on degraded pasture even if the labour were available.)

There are many species suitable for growing on vegetative strips. They can be planted purely to reduce water and soil loss, or they can be a product in themselves. In the survey, the species was not specified. Vetiver grass (*Vetiver zizanoides* L.) has been a popular choice for strips

Box 8.6 Coping with slopes

Not everyone thought that soil erosion was not a problem, or only affected other people. One of the farmers in the Muriaé valley reported that 30 per cent of his land was affected by erosion; asked about it as a constraint to his farm, he wanted to rate it at 25 out of 10. His farm was quite small, with 8ha. 'If you have cattle together in a small space, erosion can happen,' he remarked.

He was in the process of planting Angola grass (*Brachiaria mutica*) and was planting on the contour – but had not heard of the practice of using vegetative strips, and thought it might be helpful. Space might have been a problem; he was one of the few in the area who rated farm size as a serious constraint, and was not sure if he would have had space for leguminous trees, for example. But he had a hectare of sugar cane.

This might have been a farm where vegetative strips could be an option. The farmer understood the importance of contour planting, and practised it; with a small farm on sloping terrain, he had the incentive to combat erosion; and he was already growing a fodder crop that might be suitable for the strips. Space might have been a factor, but because the strips could also have produced fodder, it might not matter so much that it took up grazing space. Last but not least, unlike many farmers in the area, he had adequate labour and did not regard it as a serious constraint.

on slopes, although the germplasm can be scarce in Brazil (Dubois and Lamego, 1998, pp5–6). But plants that would be used for intercropping on flat ground may serve this purpose as well. Armando (2002, p14) suggests a number of species, some of which would be familiar to the farmers in Noroeste Fluminense and the Zona da Mata; these include pigeonpea and Napier grass (*Pennisetum purpureum*). Armando also mentions sugar cane. As stated above, sugar cane had been grown on the contour in the Zona da Mata 10 years earlier, and although it was no longer a commercial crop, almost all the farmers still grew a little, mainly for fodder; they had an average of 0.7ha each, while the Noroeste Fluminense farmers had nearly 1.3ha each. There seems no reason why this should not be planted on the contour as a vegetative barrier and used for fodder, apart from the labour requirement.

Reduction in soil loss through use of vegetative strips is hard to quantify. In the Claveria initiative in the Philippines described by Fujisaka et al (1994, pp20–22), contour hedgerows reduced soil losses from about 200 to about 20t/ha per year. However, these figures would be highly dependent on rainfall, slope, soil type, type of barrier, land-use history and other factors, and predicting the reduction in SOC mineralized during translocation would be very difficult.

Some farmers did practise contour planting and they regarded it as a good idea, rating it at 9.2 out of 10. However, constraints included labour (the farmers rated this at 6.1 out of 10). This was because tractors cannot be safely used along the contour, and expensive manual labour would be needed instead.

Of the other constraints, lack of technical knowledge was perceived as a moderate constraint. In fact, marking the contour is not difficult using an A-frame. It is a bit laborious, but farmers may be able to detect the line of the contour fairly accurately without it. Stark et al (2000, pp8–9), working with farmers at Claveria in the Philippines, found that most farmers had done this to save labour, and were usually not far out.

The impression is that farmers like contour planting and would adopt it more if the constraints could be overcome; and overall, the ratings for constraints were not very high. As Figure 8.11 shows, they were quite close to the means for all practices. Among the constraints, labour is about equal first with initial cost at 5.8, suggesting that one is a proxy for the other. 'Displaces more profitable activities' was close behind (5.7). Disaggregating the ratings showed a discrepancy between the Zona da

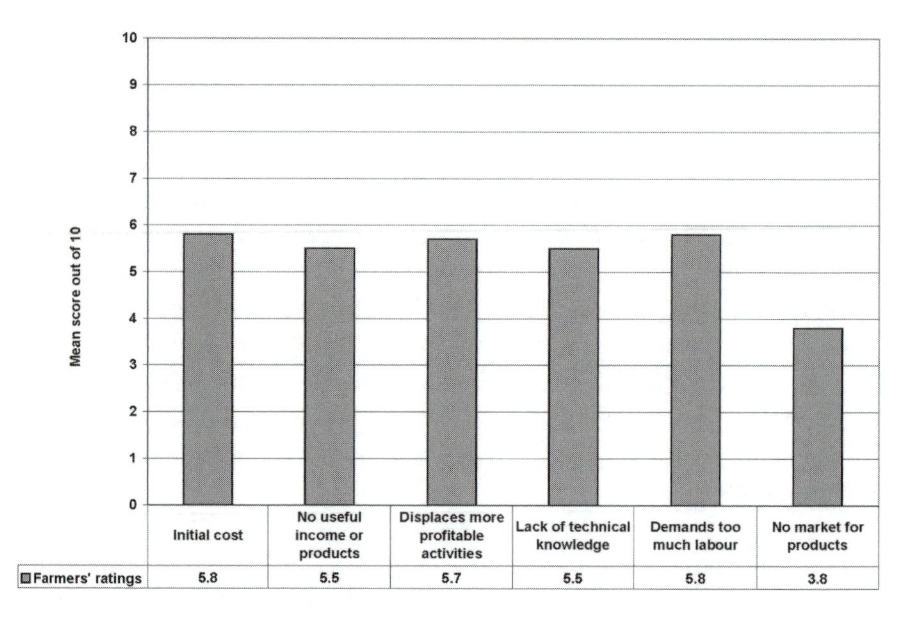

Figure 8.11 *Farmers' ratings of constraints to contour planting*

Mata and Noroeste Fluminense that was broadly similar to the overall rating for labour as a constraint. There does seem to be a pattern with vegetative strips, even more with contour planting: the activity is seen as displacing other activities because of its demand for labour, rather than for land. Contour planting would of course not demand land, only labour. Vegetative strips *would* demand some land. However, the match to the overall labour rating suggests that it was this and not land area that was the problem.

It seems that vegetative strips should be part of any carbon conservation programme, but should only be advocated on smaller plots where family labour is available, and not imposed on farmers with extensive, degraded pasture systems, even though the latter would have greater theoretical potential for losing SOC. That said, the relationship between vegetative strips and labour is variable, simply because the technology can be applied in so many ways. It will demand much labour if one is planting sugar cane along the contour. It will demand a lot less if natural vegetative strips (NVS) are used.

Stark et al (2000, pp2–3), besides describing how farmers at Claveria in the Philippines used informal methods to find the contours, also describe how instead of planting hedgerows on them, they simply left strips

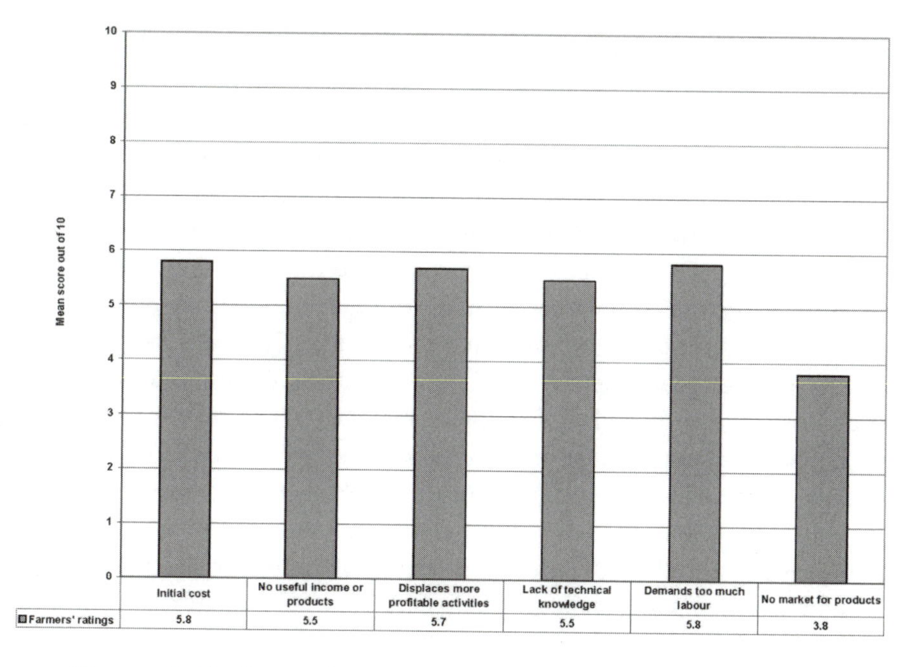

Figure 8.12 *Farmers' ratings of constraints to sustainable rotations*

unploughed; these were soon covered with 'naturally occurring grass and herbs'.

Another technique might simply be to make indentations on sloping ground that will catch water and nutrients. Gomes et al (2003) describe a series of techniques used to rehabilitate springs that help supply the city of Viçosa, not far from the Zona da Mata site in Minas Gerais. These included the digging of simple trenches and basins on steep slopes, using animal traction; the trenches were planted along the contour, and ridged at intervals. These proved effective, with other measures, at slowing runoff (Gomes et al, 2003).

Elsewhere, a range of factors has affected experience with vegetative strips. Fujisaka et al (1994, p19) asked farmers in the Claveria project what they liked about the contour hedgerows a few years after their establishment. Almost all thought they were controlling soil erosion, and nearly half reported that they provided fodder; smaller numbers cited the supply of green manure, and the more efficient use of inorganic fertilizer (because it was not lost downslope). However, about a third also reported that they competed with the crop, and other problems included neighbour's animals grazing on them and destroying them in the dry season. (In the case of the

Brazilian farmers, neighbours' cattle would probably not be a problem, but their own might be; however, only one mentioned this.)

Typically, vegetative strips could be helpful in the Zona da Mata and Noroeste Fluminense when the farmers wish to use expensive fertilizer in pasture improvement, and want to maximize benefit from it by growing maize or sugar cane to provide fodder in the dry season. It is also worth noting that vegetative strips can help control pollution, as they slow runoff of agrotoxins into watercourses. It is not clear how widely appreciated this is. In a publication of the University of California, Grismer et al (2006, p1) advocate the use of vegetative strips for this purpose with orchards, vineyards and row crops, 'especially those that are managed with bare soil between tree or vine rows'. The farmers in the survey area did not have vineyards, and their fruit trees, such as they were, tended to be clustered around the houses in the valley bottoms; but they did have concerns expressed about agrotoxins, especially in the Zona da Mata. The point is worth making, and could help carbon-related measures synthesize with other goals.

Any project funding for contour strips should be designed to accommodate different approaches to their use, and should be additional to less labour-intensive practices such as pasture improvement. Nonetheless it seems useful to have this technique in the 'basket of technologies' and to offer farmers help in meeting the costs of strips, where they fit into the farming system.

Fruit trees

Because trees of any sort sequester carbon as above-ground biomass, fruit trees are worth considering as an option, given that they produce more for both home consumption and the market than simply retaining trees for their own sake. However, there is a problem establishing how much carbon could be sequestered. According to FAO (2006, p2), non-forest land with tree coverage is 'a considerable carbon sink' in many developing countries, but it adds that data are weak.

The only systematic approach to fruit trees and the carbon market appears to be that of Kerckhoffs and Reid (2007), who prepared their report for Horticulture New Zealand, and are therefore discussing very different agroecosystems. However, their report included fruits such as plum, avocado, lemon and orange that are grown by farmers in the project area (oranges, in particular, were very common). Unlike the New Zealand crops, they are on small areas and for family consumption; the varieties would also not be the same. But Kerckhoffs and Reid's figures may serve as a rough guide.

Kerckhoffs and Reid assumed that the dry weight of a tree would be approximately 50 per cent carbon and then multiplied this by 3.67 to establish the amount of CO_2 that would be sequestered. On this basis, they think that 'even the most vigorous stands' of orchard fruit trees were unlikely to sequester much more than 50t/ha of CO_2, or about 13.6t/ha of plant organic carbon (Kerckhoffs and Reid, 2007, p3).

In terms of above-ground biomass, this is clearly superior to pasture or cropping, but gives no indication of total system carbon, including the amount of carbon sequestered in the soil of the orchards, or in the sward between trees. Without these figures, it is not clear what the climate-mitigation potential of growing more fruit trees would be. Kerckhoffs and Reid say that, in the time available, they were not able to consider these carbon pools. Given the wide variation in soil carbon that could result from different management practices in orchards, from both sward and possible erosion through its absence, this is very important. If the sward between trees is removed to prevent it from competing with them for water and nutrients, significant erosion may result. For example, a few years ago, the Ministry of Agriculture in Spain estimated that as much as 80 million t/ha per year of soil were being lost from the 1 million hectares of olive plantations in Andalusia alone (Pohl, 2001, p3).

Moreover the life expectancy of an orchard will be 10–30 years, depending on species, after which it may be clear-felled; and how this is done, and what happens to the dead wood, clearly affects the net effect growing the orchard has had on radiative forcing (Kerckhoff and Reid, 2007, p3). This could be a tricky question; if it is used for fuelwood, then it could wipe out the mitigation effect – but possibly not if it was burned in place of native forests, or replaced some other source for people who would have burned wood anyway, as Zona da Mata farmers would. Overall, though, it seems unwise to look at fruit trees as an important tactic for sequestering carbon. However, they could be positive if replacing degraded, as opposed to well-maintained, pasture (although, as we have seen, pasture rehabilitation or crop rotations might do this just as well).

The farmers surveyed were quite positive about fruit trees, seeing them as good for the soil (not least because of their litter, as one farmer pointed

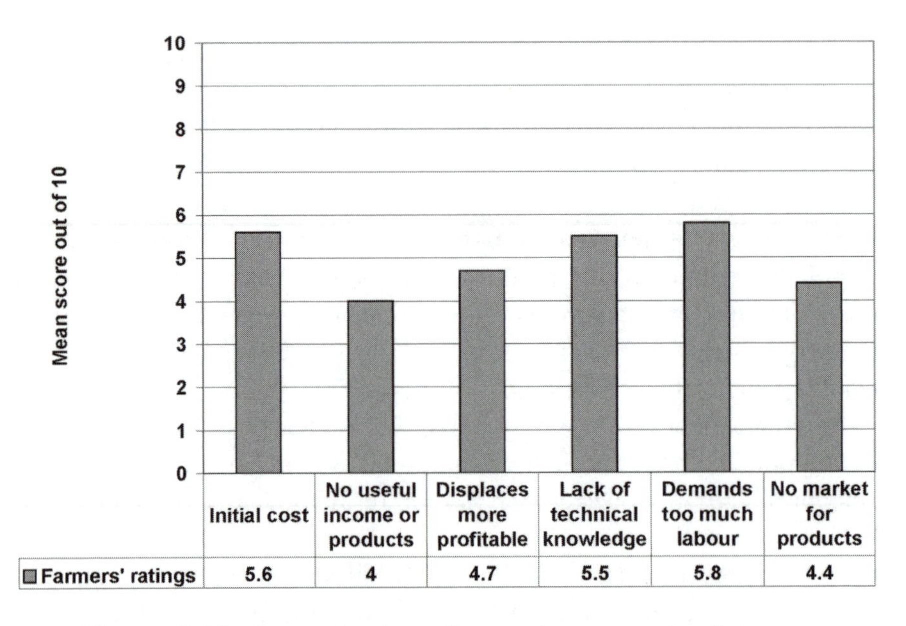

Figure 8.13 *Farmers' ratings of constraints to growing fruit trees*

out). However, few wanted to grow any more of them. This was not because fruit trees were especially difficult to grow. Everyone already did have a few, and the ratings for constraints (Figure 8.13) were no higher than those for most practices. Neither can marketing be the explanation; true, the Noroeste Fluminense farmers had the state fruit scheme, Frutificar (see Box 8.7), to help with this, but the Zona da Mata farmers too did not rate marketing very high.

As virtually none of the farmers in any of the locations had much to say about fruit trees, with one or two exceptions (see Box 8.7), no clear explanation emerges. It is probable that farmers just did not see their small household fruit plots as anything related to a commercial crop. (Although one in the Zona da Mata did say, intriguingly, that he had heard fruit trees could 'clean the air', and he would like to know more.)

There is one possible alternative explanation: that it was in the main women who tended the fruit, whereas men looked after the cash crops. As the interviewees were mainly (though not always) men, this may be a gendered response. Indeed the gender division, and the way it affects participation in land-use mitigation projects, has been noticed by Boyd (2002, p74) in her discussion of the Noel Kempff project in Bolivia, and by Corbera (2005, pp51–52) regarding the Fondo Bioclimatico project in Chiapas, Mexico. In the latter case, Corbera reported that women were

Box 8.7 Going for guava

In the Muriaé Valley, one of the farmers had tried to branch out into fruit trees. He had obtained support from the state. In 2000, Rio started the Frutificar programme (it is not an acronym; *frutificar* is Portuguese for to make fruitful). The objective was to intensify agriculture in the state by giving loans and marketing guarantees for areas of up to 3ha of fruit for farmers who had been in place for at least five years. The state reports that it has been a success, directly or indirectly creating 20,000 jobs in the state. This farmer – who had a 65ha farm, most of it pasture – was not happy; the programme had provided too little support for disease, he said, especially in passion fruit, and only about 20 per cent of the farmers got their investment back. He had also invested in a processing plant for guava, he said, but had closed it because it wasn't profitable, and was left with 120 trees.

In the Zona da Mata, by contrast, a farmer with 7ha had been concentrating for the previous two years on guava (not on the whole area; he had one or two other fruits, plus 1ha of pasture). He also had a small processing unit and was selling the finished product. He was not without problems – water was short, and fertilizer was expensive at Rs45 (about £11.80) a sack. But he was reforesting part of the farm with 3000 eucalyptus trees, and supplemented imported fertilizer with pig and chicken manure. He seemed to be doing well.

more involved early in the project, when it had more development-related objectives; at that stage, they discussed planting fruit trees in their household gardens (and improving their cooking stoves). As the project shifted emphasis to carbon accounting and tree planting, however, their suggestions received less attention and they took less part in the project. For the survey described in these pages, there were neither the resources nor a suitable methodology for assessing gender issues, but they should be approached more carefully in any actual project plan; indeed Boyd (2002) actually looks at the Noel Kempff project through such a framework.

Despite these unknowns, fruit trees as a sequestration practice are not obviously better than others – in particular, pasture improvement, fertilizer use and perhaps organic agriculture – where the farmers' response was more consistently favourable, the constraints clearer and the sequestration potential better, and where there would be no new marketing issues to deal with, as the nature of the product would not change.

Fruit trees were the last of the options discussed with farmers as part of the survey. Farmers' responses regarding the individual practices should have provided insights, first, into how to work with farmers to increase terrestrial carbon; and second, the way in which the broader economic environment defines farmers' use of that carbon. These questions are examined in the next chapter.

Notes

1 Cited in text as IBGE (1996). No page numbers will be given; the survey does exist as a printed document, both for Brazil and in separate volumes for the individual States, but these would not be widely available outside Brazil and the author has mostly used the IBGE's online tables.

2 Like Kotschi and Müller-Sämann, whose document was published by IFOAM, Edmeades was associated with a body; in his case, Fertilizer Information Services Ltd. of Hamilton, New Zealand. In both cases, the affiliations are clearly stated in the papers.

3 In Portuguese, *plantio direto na palha* – this is the phrase often used in Brazil, although in the survey it was referred to as *cultivo minimo*, as this covers all forms of reduced tillage.

4 Federacão Brasileira de Plantio Direto na Palha (Brazilian Federation for Direct Drill through Straw).

9

The Heretic's View

> *In the face of immediate problems of poverty, food insecurity and poor agricultural productivity, soil degradation may be readily relegated down [farmers'] list of priorities. How can resource-constrained farmers be expected to adopt practices that in the long term may improve production, but in the short term realise no net benefits, or even net losses?* (Giller et al, 2009)

The objective of this case study was to see whether there were technologies for carbon sequestration that would fit into the farming system in the Atlantic Forest region, what they were, and what might be learned from this exercise that could be applied to sinks projects in agriculture generally. The hypothesis advanced at the end of Chapter 7 was that there would be a need for a 'basket of technologies' so that farmers could pick, choose and if necessary modify technologies according to the nature of their particular farm. This 'basket' hypothesis recognizes the farm-specific nature of agricultural practices, and accords as much importance to stimulating the farmers' own processes of innovation as to simple technology transfer. It is suggested by many years' research into technology adoption and adaptation in general, culminating in Sumberg and Okali's (1997) conclusions to this effect. It might also partially address the equity concerns raised by Tschakert (2004, 2006, 2007), although it would not eliminate them.

The case study in the Atlantic Forest suggests that this approach is correct for carbon sequestration. Prospects for adoption were highly dependent on labour availability, topography and access to technical advice, and these varied between survey sites and sometimes between neighbouring farms. Contour planting and CT were good examples of this; there was limited enthusiasm for them for pasture improvement, but farmers with sloping land felt differently. So someone with (say) steeply sloping land but not much labour might decide to create vegetative strips but do so simply by making indentations in the ground or ridge-tilling on the contour, leaving natural vegetative strips (NVS) to grow.

This is important because carbon sequestration in agriculture might be based, not on measurement of carbon sequestered, but on payments for adoption of specific practices. However, the 'basket' hypothesis suggests that a farmer might not adopt the technology if it is rigidly defined. Left to adapt it to his or her specific needs, s/he might. The survey seemed to confirm this. But accommodating a more liberal approach will cause serious accounting problems.

It will be hard to circumvent this. The author does not believe production of CERs in project-based sinks are impossible in agriculture. However, this book has shown how difficult they will be in any but the most homogeneous farming systems, both because of farmer-specificity but also because of the methodological constraints discussed in Chapter 5. The solution may lie in sectoral programmes covering a whole region or state, with the carbon budget produced through a combination of remote sensing and ground-truthing over periods of many years. A sectoral programme – perhaps for a region, rather than a whole country – would reduce per-tonne transaction costs and allow a broader mixture of mitigation and verification approaches. The other alternative is the use of payment for ecosystem services (PES) that does not lead to the issue of CERs, discussed in Chapter 5; this has great potential, especially if REDD+ lives up to its funding promise, but the non-issue of tradeable credits will always restrict the available funding base.

Another topic arose often in this chapter – scientific uncertainty – and it has been very important; but, again, it relates to broader themes of this book, and will be discussed in the final chapter. At this stage, it may be useful to look at two themes that arise specifically from the case study.

The first concerns the technologies that the farmers did and did not want to adopt, as there is a dissonance between their preferences and those envisaged by the GEF input in Noroeste Fluminense. This may point to major distinctions between different philosophies in agricultural development – distinctions that may have to be abandoned if sinks projects are to succeed in agriculture. This is discussed in the next section.

The section that follows it looks at the second theme – that the economic situation of farmers in the Atlantic Forest, and particularly in Noroeste Fluminense, makes it very difficult to exploit the enormous biophysical potential for sequestering carbon on their farms. The author believes this may point to a flaw in the way in which PES concepts are applied to agriculture in isolation from the broader economy in which it operates.

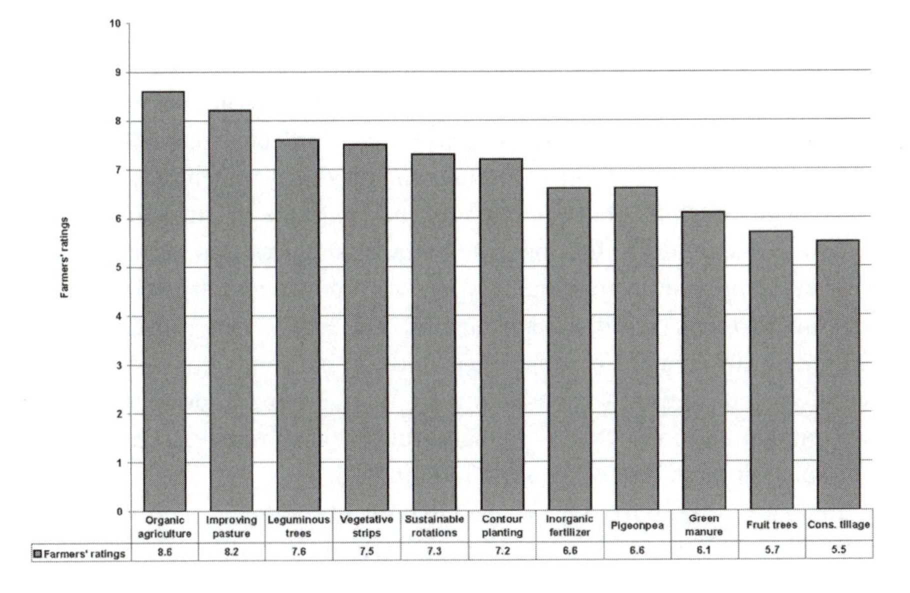

Figure 9.1 *Farmers' willingness to adopt (would like to use/use more)*

Perhaps the farmers are being asked to internalize externalities that should really be borne by those who use their products.

Agricultural Ideologies: A Barrier to Adoption?

Figure 9.1 ranks the carbon-friendly practices according to farmer preference as expressed in answer to whether they would like to either use a practice or, if they were already doing so, use it more. The gaps between the ratings are not that large; for example, that between the top practice, organic agriculture, at 8.6 and pasture improvement at 8.2 was not significant. But there is a significant difference between the first-rated practice, organic agriculture at 8.6, and the one that farmers were least keen to adopt, which was conservation tillage (CT) at 5.5.

As stated in Chapter 6, the practices for the case study were selected partly with reference to those planned for the GEF-assisted Rio Rural project in Noroeste Fluminense. At this stage, it is worth comparing those options with those the farmers thought they might want to adopt. The project document (GEF, 2005b) lists two sets of practices: those that would be part of the baseline project as implemented by the State of Rio de Janeiro without GEF funding, and those that would comprise the 'GEF alternative'. The way GEF operated was described in Chapter 5; to recap briefly, it does not mount its own projects but provides funding for existing development proposals in order to make them more environment-friendly. This is known as incremental funding, and produces incremental benefits; thus a state development project might fund agricultural intensification of some kind, and GEF might provide extra funding for that activity on the understanding that it should be modified to produce global environmental benefits not envisaged in the baseline project. So the incremental funding might add an element for (say) CT, and the carbon sequestered would be calculated, in cash terms, as an incremental benefit along with other environmental goods such as reduced runoff of agrotoxins. This is what was being done in Norte/Noroeste Fluminense.

Table 9.1 is taken from the project document for the GEF project in that area (GEF, 2005b, p52) and shows the Rio Rural and GEF components for one sector of the project, rehabilitation of degraded land – which is where some of the best sequestration potential would be. Figure 9.1 suggests that some of the activities under GEF funding will be welcomed by farmers; for example, contour planting – they lacked the resources for this, but understood its value. Other activities look problematic. The farmers were not at all keen on minimum tillage. This is striking, as globally it is probably the management practice most frequently advocated for carbon sequestration in agriculture; moreover it is widespread in other parts of Brazil, which is – along with the US – the world leader in applying this technology. Their reluctance sprang partly from unfamiliarity; few had considered it before. But it was also less applicable to their farms than to those in the southern states, where there is greater emphasis on arable farming.

They also had doubts about green manure. They were not against it, and the most important single constraint, technical knowledge, might be addressed through extension funded by GEF. But labour was also an important constraint. While it might be alleviated temporarily by project funding (assuming the labour was available – a topic discussed in the next subsection), the constraint would return afterwards. As for contour planting, many thought this a good idea but few could do it because it was dangerous with a tractor, and very expensive if done by hand. At first sight, then, it might appear that the proposed GEF activities are imposing 'ecologically correct' activities on farmers who do not really want them.

Table 9.1 *State- and GEF-funded components for rehabilitation of degraded land, Rio Rural project*

State funding	***GEF***
Use of fertilizers and corrective measures for soil-fertility restoration	Organic manure Green manure Minimum and zero tillage
Erosion control on rural roads	Grass bunds Contour cropping Terrace planting Crop rotation Cover crops
Conventional machinery and equipment for use in conventional productive systems	Equipment for soil conservation and management
Soil preparation practices (ploughing in terraces)	Mechanical erosion control

However, that would not really be fair to the GEF project, as the ratings do not tell the whole story. Organic agriculture demonstrates this. The farmers actually rated inorganic fertilizer higher than green manure, at 6.6 against 6.1. But because the Zona da Mata farmers did not wish to give a rating for inorganic fertilizer, the sample size makes this difference insignificant. In any case, although the farmers were not very enthusiastic about green manure, they were a lot more so about organic agriculture. They had good reason to be, given their concerns about agrotoxins;[1] according to one national database on poisons, SINITOX (Sistema Nacionalde Informação Tóxico-Farmacológica), of over 34,000 occupational poisoning incidents reported between 1996 and 2001, nearly a third (32.9 per cent) involved agrotoxins (Da Silva et al, 2005, p899). Moreover agrotoxins have a financial as well as health cost; Brazil consumes some US$2.5 billion worth of them a year, a sizeable chunk of the US$20 billion worth sold world-wide by the 20 major producers (op. cit., p895). Neither should it be assumed that harm can arise only from pesticides and herbicides. As Neves et al (2002, p313) point out, ingestion of nitrates and their transformation into nitrites is also a serious health risk.

But there are dangers in assuming, as GEF does, that organic or agro-ecological approaches will necessarily provide incremental environmental benefits over conventional approaches. For a start, the assumption that organic agriculture has more potential for climate-change mitigation than conventional agriculture is open to dispute; as argued in the previous chapter, it should in principle, but in practice it is far from clear. Besides, the agroecological approach will be appropriate for some farms but not others.

A perspective on organics from within the Brazilian agricultural research system – indeed, within Rio de Janeiro State – is given by Feiden et al (2002), who say that the price premium for organic products has attracted entrepreneurs 'who envisage only an immediate profit, without many environmental concerns'; this market pressure, they say, encourages 'organic' production systems that are actually based on conventional, technological forms of production – for example 'organic' monocultures (op. cit., p183). Moreover, they are concerned that the process of conversion or transition to organic agriculture is often regarded simply as a period of 'quarantine' in which chemicals may not be used (officially defined as 12 months for annual vegetables and perennial pasture, and 18 months for perennial vegetables). In fact, they say, the transition should not be regarded as a simple quarantine but as a time for reorganization and 'consolidation and maturing of new knowledge, accompanied by an active realignment [*ressituação*] of farmers and the environment' (op. cit., p185). The authors go on to outline some of the principles they consider to be fundamental to an agroecological approach, including maintenance of organic matter and soil biota, management of soil fertility, creation of biodiversity to maximize the ecological services provided by the system, and – perhaps most important – the replacement of fertilizers with greater optimization of nutrient cycles and biological systems of fertility control (op. cit., p192).

Is this too prescriptive? Although many farmers said they were interested in organic agriculture, it seemed many regarded organic agriculture as simply not using chemicals. Not all had the space or labour for green manure, and had they seen organics as implying such an upheaval in the long term, some might have been less keen, especially in Noroeste Fluminense. There is no bar to farmers obtaining organic certification simply through non-use of chemical inputs; done that way, organic farming can be seen simply as a business proposition rather than a philosophy.

However, attempting to implement organic agriculture *without* seeing it as a totality could result in a loss of fertility through not using chemical inputs, but not replacing them properly. This would also reduce on-farm carbon stocks. Arguably this would make little difference in the Zona da Mata, where farmers already use little chemical fertilizer; they might as well get certified organic and charge more for their produce. But little carbon would be sequestered this way. As for the prices, more might be achieved simply by taking their own produce to market; as the extension agent in Piraúba pointed out, much of the profit is going to middlemen. So although organics should not be dismissed from the farmers' point of view, within the context of the survey – seeking options to sequester carbon – organic agriculture only makes sense if applied properly.

In the Atlantic Forest, therefore, it might be argued that the best way of sequestering carbon into degraded, nitrogen-deficient pasture is actually to use as much inorganic fertilizer as possible – and some farmers in Noroeste Fluminense (although not in the Zona da Mata) did want to use more. Vlek et al (2004), in an argument referred to briefly several times already in this book, go farther, suggesting that more efficient farming could sequester carbon by allowing the conservation and possible regeneration of native ecosystems. They propose that this be done through the greater use of fertilizer. Its potential for increasing production, they argue, has nowhere near been reached in some regions:

> *Assuming that 20 percent more fertilizer use in developing countries leads to 22.9 million ha recoverable land that can be used for re-vegetation, we obtain a potential carbon sequestration between 25.1 and 59.4 Mt yr^{-1} for all the regions combined, depending on the assumed sequestration rate, with an average of 42.3 Mt yr^{-1}.* (Vlek et al, 2004, p226.)

There are some problems with this, not least that extra yields from fertilizer input are calculated in part from pre-Green Revolution yields, when, as they say, fertilizer inputs were 'presumably insignificant' (Vlek et al, 2004, p225); but also, other inputs would also have been lacking and populations, labour inputs and farm-gate prices would all have been different. More important perhaps is the assumption that increasing the efficiency of existing farmland would necessarily reduce incursions into native vegetation or increase re-vegetation of former forests. It might do the opposite. In the case of the Brazilian Amazon, both sides of this argument have been proposed (see for example the discussion of REDD+ in Chapter 5 above).

Even so, the use of inorganic fertilizer to rehabilitate degraded pasture does have potential in the Atlantic Forest region and could result in restoring SOC to something quite close to its level under native vegetation, while retaining the land as a farming asset. Yet this solution, although favoured by the Noroeste Fluminense farmers, was not attractive to those in the Zona da Mata. So there are two distinct approaches – agroecological, and input-based – that could be used to sequester carbon in the region. Both would be appropriate for at least some farmers, but neither would cover everyone's needs; and yet they seem to be mutually exclusive.

But perhaps they are not. Crosson and Anderson (2002), in a discussion paper for the International Food Policy Research Institute (IFPRI), compare two different philosophies of agricultural development, which they call the 'conventional' and 'alternative' systems. The first they define as relying on technological inputs such as high-yielding cultivars and purchased inputs, in particular inorganic fertilizers and pesticides; the second seeks to minimize purchased inputs, relying instead on rotations, mixed cropping and crop–livestock integration (Crosson and Anderson, 2002, p3). Proponents of alternative systems, they say, are not indifferent to yields but do not regard them as the prime objective. They are also more management-intensive.

There is a further distinction. 'Some advocates of the alternative system like it also because, in their judgement, it promotes a more equitable distribution of income and political power in the countryside,' add Crosson and Anderson (ibid.). Although they do not say so, that implies that a commitment to 'alternative' or agroecological approaches may be partly ideological. But perhaps an adherence to technology-driven, input-intensive agriculture is scarcely less so, implying a technocratic and managerial view of the natural world.

If ideology does drive the debate between these philosophies, then it may have more to do with researchers' prejudices than farmers' welfare – or the carbon sink. It might be helpful to discard such prejudices and find more utilitarian drivers for policymaking. Crosson and Anderson suggest that the two approaches have more in common than they seem to, because the literature on both sides stresses the importance of research. 'This suggests that scientists in agricultural research institutions will find themselves following paths that combine the best features of each system,' they state (Crosson and Anderson, 2002, pp28–29).

This does not seem assured when opinions are strongly held. However, there are practical examples that illustrate the futility of the ideological divide. Garrity (1999, p334) quotes the example of the natural vegetative strips (NVS) at Claveria in the Philippines, described in the previous chapter: 'But where do natural vegetative strips fall on the spectrum of "conventional" versus "alternative" agriculture? They ... are an embodiment of the application of agroecological principles. ... But [they] also ... reinforce conventional approaches, as they tend to indirectly stimulate the use of commercial fertilizers and commercial cultivars' (Garrity, 1999, p334).

This is the type of technology suggested by the mixed responses in the survey. Insisting on a purely agroecological approach on the grounds that it is 'better for the environment' may result in complete non-adoption of practices that provide environmental services; or, worse, attempts to farm without chemical inputs on the grounds that this is 'organic', without maintaining soil fertility in the manner that proper organic farming demands. Yet if access is not provided to alternative systems, farmers will remain tied to external inputs that they cannot afford and that have serious health and sustainability implications. Besides, response to fertilizer will be dependent on existing organic matter, and the prospects for using fertilizer to increase SOC are much better if it is used in conjunction with organic inputs such as crop residues – a point emphasized by Bationo et al (2007, p18) with reference to the Sahel. Indeed the use of inorganic fertilizer on its own could cause a *loss* of SOC (Bationo et al, 2007, p20). So a rigid adherence to either modernist or agroecological approaches will mean that not much carbon will get sequestered.

One way of bridging the gap may be to take a more liberal view of fertilizer. Although, as stated earlier, the IFOAM definition of organic agriculture excludes urea, the official Brazilian definition does not explicitly do so, requiring simply 'the elimination of agrotoxins and other toxic artificial inputs'.[2] Urquiaga and Malavolta (2002) argue that urea, although synthetic, is 100 per cent organic and say that it can 'contribute significantly to the development of organic agriculture in the country, especially in those areas where the availability of animal manure and cultivation of legumes as green manure are very limited or not very viable' (Urquiaga and Malavolta, 2002, p333). They suggest applications that would be especially suitable, including fertirrigation; also use in composting crop residues low in nitrogen, which they think would be an excellent solution to shortage of animal manure.

This approach would constitute something approaching integrated nutrient management (INM). Put simply, INM is the best of both worlds. Gruhn et al (2000) define it as follows: 'INM's goal is to integrate the use of all natural and man-made sources of plant nutrients, so that crop productivity increases in an efficient and environmentally benign manner, without sacrificing soil productivity of future generations' (Gruhn et al, 2000, p15). This sounds, and in principle is, straightforward; farmers must use the most productive and sustainable combination of organic and inorganic inputs.

However, farmers still have, as individuals, to make the trade-off between productivity and sustainability; and they have to find the best balance of inputs for their particular farm: '[F]armers, with the aid of extension services, have to be given access to and choose the most appropriate and cost-effective technologies for their particular circumstances. Farmers also need to participate in the development of these technologies and become knowledgeable about managing soil fertility ...' (Gruhn et al, 2000, p20).

This change of approach also accords with the 'basket of technologies' theory; carbon sequestration initiatives, like those for broader agricultural development, should give farmers resources to reach a goal but leave it to them to decide how to do it. The GEF alternative could perhaps include some subsidized urea, for use on maize, say, or on the limited sugar cane grown for fodder. At the same time, however, it could provide assistance with biological nitrogen fixation through green manures, forage legumes and leguminous trees. It could also finance activities whose benefits would be maintained when funding was no longer available, such as vegetative strips, laying less emphasis on those like contour planting, which, despite their importance, might be abandoned later.

Meanwhile, carbon sequestration programmes in general should not be designed by those on either side of the ideological divide.

Internalizing Environmental Costs: Who Should Do It?

The second conclusion suggested by the Atlantic Forest case study arises from the constraints to adoption of practices. There are ratings here which raise questions concerning any form of payment for environmental services (PES) in the region.

The highest-rated constraint is initial cost. It is rated significantly higher than the next constraint, lack of technical knowledge, and labour, which is just below. The lowest constraint, marketing, is rated significantly below the next up, 'No useful income or products'. The steps down between the practices in between were not statistically significant. It seems reasonable to say that initial cost was the most serious constraint but that the others were also mostly worth addressing. The ratings also seem compatible with those given for overall constraints to the farming system, which put credit (which is a rough proxy for initial cost) at the top and labour in roughly the same place, with marketing quite a long way down.

However, these ratings may be misleading in some respects. Farmers interpreted 'market for products' as meaning that they could or could not sell their milk or produce. The price they got for them was a different matter. 'Displaces more profitable activities' is also misleading. It was not always answered on the basis that space taken up by one activity would exclude another (although the Seropédica market gardeners, with their small plots, did take it that way). It seems also to have been a proxy for labour in the case of the Noroeste Fluminense farmers, as the gap between their ratings and that of the Zona da Mata farmers for this constraint coincided at least twice with their differing overall ratings for labour,

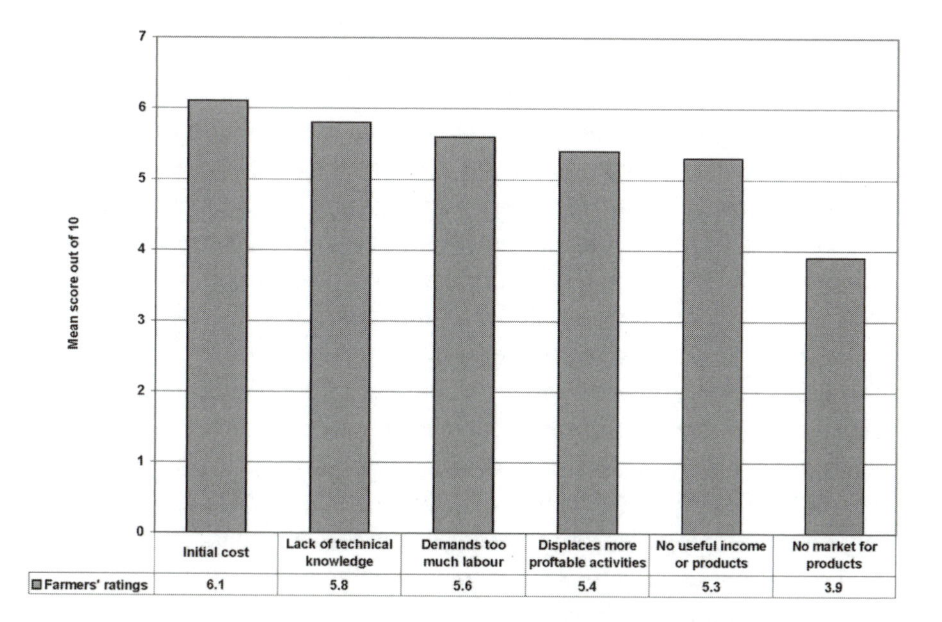

Figure 9.2 *Mean ratings for constraints, all practices*

suggesting that the displacement was seen as one of labour and not of space. There also seems no other reason why they would give higher mean ratings for this constraint across all practices, when they had bigger farms.

The same may apply to initial cost, as the Noroeste Fluminense farmers marked this higher than those of both Seropédica and the Zona da Mata; 7.8 against 5.3 and 6 respectively, and the gap was significant. Yet their overall rating for credit as a constraint was lower than that of the Zona da Mata farmers, although the difference was not significant. Once again, there seems to be a link with labour. (Although this was not the case with fertilizer; here, 'initial cost' did seem simply to mean that the input was expensive.)

It would be wrong to overemphasize this. The Zona da Mata farmers, with their family labour, were not so constrained in this way; they were more worried by access to technical help. Besides, there is a genuine problem in all locations with access to credit. Even so, it seems it is often labour availability that will define whether a farmer can adopt a practice or not.

Why is the labour not available? The obvious answer is rural–urban migration and as Table 9.2 shows, this is almost certainly part of the answer. But there is evidence that much of the labour lost from agriculture over the last 30 years has not, in fact, gone far. Table 9.2 gives the figures for the economically active rural population. The original from which it has been adapted, in Da Silva et al (2002, p44), gives corresponding figures for urban areas as well; they are omitted here, but the overall economically active urban population increased by a massive 43 million during the period covered, a growth rate of 2.6 per cent for the first period and 1.7 per cent for the second. However, the rural population has not declined by enough to fuel more than a fraction of that growth, and in the 1990s it actually grew, albeit slowly. The urban growth has come from elsewhere. So it is hard to see how rural–urban migration has stripped available labour from the farming sector. Yet something has.

The answer lies in the growth of non-agricultural rural occupations, referred to by Da Silva et al by their Portuguese acronym of ORNAs,[3] which increased by 3.7 per cent in the 1990s while agricultural employment decreased by 1.7 per cent. This suggests that ORNAs have been not only taking employees from farming, but have also soaked up much of any population increase as well. If this is so, the implication is that many agricultural employees have found better-paid jobs outside farming but without joining the migration to unplanned and unserviced settlements in the cities – which can only be good. If farming has suffered as a result, then perhaps it should have paid higher wages.

In fact, this is not what has been happening. Among the new jobs, the most numerous seem to be in domestic service; the next category is civil construction. 'It is notable that both the sectors are known for the low education levels and professional qualifications of the majority of their workers,' say Da Silva et al (2002, p51). This is reflected in low rates of pay (Da Silva and Del Grossi, 2001, p451). Moreover, between 1992 and 1999 rural unemployment in Brazil grew by no less than 10 per cent, according to Da Silva et al (2002, p63), who also identify a rapid growth in the late 1990s of families that have neither jobs nor land. 'The great bulk of the rural population engaged in RNFE [rural non-farm employment],' say Da Silva and Del Grossi (2001, p447), 'are working in poorly paid jobs that demand few skills and little schooling, even in the most developed regions of the country.'

There is evidence that all this is not just a Brazilian phenomenon. Reardon et al (2001, pp395–396) reviewed studies in 11 Latin American and Caribbean (LAC) countries and reported that non-farm income accounted for about 40 per cent of LAC rural incomes. They too found that 'poor households and poor zones often lack access to the better-paying nonfarm employment … [and are pursuing] RNF [rural non-farm] activities that are the equivalent of "subsistence farming"' (Reardon et al, 2001, p396).

This has been accompanied by the increasing prominence of the Movimento dos Trabalhadores Rurais Sem Terra (MST, the Movement

Table 9.2 *Growth and decline in economically active rural population by category, Brazil, 1981–1999**

	Number of people (millions)			Annual growth or decline (%)	
	1981	1992	1999	1981–1992	1992–1999
Economically active (total)	34.5	32	32.6	−0.7	0.2
Employed (total)	13.8	14.7	14.9	0.6	−0.2
Agricultural	10.7	11.2	10.2	0.4	−1.7
Non-agricultural	3.1	3.5	4.6	1.2	3.7

Source: adapted from Da Silva et al, 2002, p44

* Excludes Northern Brazil (except State of Tocantins)

of Landless Rural Workers). Founded in 1978, this movement of landless workers aims to take over what it considers to be underused land so that it can be worked. The concentration of land in Brazil has increased; in 1966, properties below 100ha accounted for 20.4 per cent of the land and those over 1000ha were over 45.1 per cent; by 1992, this had changed to 15.4 per cent and 55.2 per cent respectively (INCRA, 1992, quoted in Meszaros, 2000, p3). This process may be connected with the exploitation of new land in the Cerrado and elsewhere rather than changes in the ownership of existing properties. Nonetheless, land distribution is skewed, and Meszaros (2000, p3) reports considerable public sympathy for the MST, claiming that at the time of a march on Brasilia in 1997, a poll for Brazil's National Confederation of Industry found that '94 per cent of respondents considered the struggle for agrarian reform to be a legitimate one; 77 per cent considered the MST to be a legitimate movement; and 88 per cent felt that unused lands should be confiscated and passed on to landless families' (ibid.). Perhaps unsurprisingly, the farmers the author interviewed in Noroeste Fluminense, where there was MST activity, were less sympathetic; they claimed that land occupied with the help of the MST was rarely worked well for long, and that sometimes the occupiers obtained title, then sold the land and moved on to the next occupation. The MST is a contentious movement, but it is certainly a symbol that something is wrong in the countryside. It is tempting to speculate that some of the MST's supporters could be settled on underexploited or degraded pasture and helped to intensify production with the use of carbon-related funds.

If the labour is in fact still there, is not highly paid, and is even often unemployed, why are the farmers complaining that they cannot employ people? The answer seems to be that they cannot pay even the poor wages demanded. Da Silva and Del Grossi (2001, p449) report figures on per capita rural incomes compiled by the Projeto Rurbano of the University of Campinas. Taking all rural incomes as 100, employees as a group earned 91 per cent of the mean, and the self-employed 88 per cent. Non-farm employees earned 134 per cent and farm employees just 61 per cent – US$58.68 at the 1997 exchange rate. Moreover the national per capita income was 260 per cent of the rural. These figures should be approached with caution regarding the survey area because – as the authors themselves make clear – 49 per cent of rural families are in the North-East of Brazil, which is a poor region. Still, that skewed concentration would have dragged down non-farm as well as farm wages as well, so these figures are indicative in relative terms. They may well be mirrored elsewhere in LAC; Reardon et al (2001, p395) also report that returns to rural non-farm employment, poor as they often are, are usually superior to agricultural work.

The implication is that farms are not being prevented from intensifying – and increasing their carbon sink – because there is no labour; they are just not making enough profit to employ it. It may be that competition from a globalized food market is partly responsible, as Cassel and Patel (2003) claim. 'Liberalization had a predictable and negative effect on prices: world prices for Brazil's major crops … have been falling since the early 1980s,' they say. 'From 1980–1991 alone, real producer prices for both domestic crops and exports were cut in half. Prices have continued to drop in the 1990s. Over the last thirty years, rice prices have declined 53 per cent and maize prices by 60 per cent' (Cassel and Patel, 2003, p1).

With commodity prices this low, there is clearly pressure to adopt extensive livestock systems that demand less labour per hectare. But such systems are themselves not very profitable and, because they also add less organic matter to the soil, they are in effect using up the farmer's natural capital, as well as creating a sub-optimal carbon sink.

Lopes et al (2004) looked at the profitability of 16 dairy farms near Lavras, not far from the Zona da Mata in Minas Gerais. Some farms were covering their operating costs but they were not, in the main, covering the long-term costs of replacing capital. 'In the economic analysis … it is concluded that dairy farming has the conditions to produce in the medium-term, and, in the long term, that the farmers are decapitalizing,' they state (Lopes et al, 2004, p891). Moreover the capital goods reviewed in their analysis, although including animals, tools, vehicles and machinery (op. cit., p884), did not appear to include natural capital in the form of soil structure, biodiversity, invasive species or any other of the factors that would affect both carbon content and long-term productivity.

It is hard to account for these types of capital, but it should surely be done. As FAO (1994, Ch. 7, p1) points out: 'Changes in natural capital are not currently included in systems of national accounting. … In the case of soils, these were treated as the "land" factor in classical economics, priced at the market value of farmland.' The problem with this is that the productive capacity of the land, including its SOM, can be used up as with any asset, including machinery. In effect, the farmers are doing just that in order to subsidize urban milk consumption.

This raises a problem with applying a payment for ecosystem services (PES) programme. What a PES does, is make the user of a resource – in this case, carbon – internalize environmental costs that would otherwise be imposed on the wider community – people in a watershed, a nearby city, say, or, as in this case, the world as a whole. If the farmers were making a profit from the pasture, it would be reasonable to ask that they internalize the long-term costs. However, in this case, who is using the resource? It seems to be the consumers, rather than the producers, of agricultural produce, because they are using the organic matter in the soil but are not paying the farmer enough for it. Should the end user not be at least partially responsible for internalizing this cost? This does not mean that there should not be a PES scheme for farmers to sequester carbon. But carbon trading might not be the only way to pay for it. For sustainable pasture improvement, a milk levy might do instead.

The implication of this is that the consumer should pay; in other words, that farm-gate prices should be kept up to a minimum level so that farmers can farm sustainably – especially in countries where prices have been kept artificially low for political reasons, causing a market distortion (which is what is clearly taking place, if there is a demand for a product but the producers are decapitalizing). This might be part of the answer, as might controls on input prices; but such measures would need great care if they were not to have perverse consequences. Pagiola (1999b, p49) points out that manipulating prices is 'a very blunt instrument. ... Unless the subsidy ... is narrowly targeted ... it is likely to affect activities far beyond its intended scope, thus imposing substantial budgetary costs and creating inefficiencies elsewhere in the economy.' There is also a risk that the subsidy could in effect come from consumers with limited incomes; a percentage of supermarkets' profits on milk sales might be more just, easier to collect, and as effective.

However, it seems fair that if the land is to be maintained as a capital asset, the cost of doing so must be reflected in farmers' place in the market. This is the second conclusion reached from the case study, along with the need for a flexible approach to technology transfer.

Working with Farmers: Some Conclusions

The previous chapter presented the results of a survey of farmers in the Atlantic Forest biome designed to establish which options for carbon sequestration were most congruent with farmers' objectives, and what they perceived to be the constraints to those practices. It was hoped that light might be shed on the best way to implement sink projects in agriculture, and whether any links were revealed between carbon and the broader economy. Before reviewing the survey data in detail, the sample was compared with the means for the region, and found to be broadly typical in farm size and status. It was also found roughly typical in the percentage of pasture used.

The survey results suggested a bias in favour of pasture improvement as the best way to sequester carbon. It also suggested that the use of more fertilizer to assist in the establishment of *Brachiaria* was a practical way of doing this. A strong extra tactic was the use of leguminous trees to reverse soil degradation, also providing other services (such as fence posts and shade for cattle).

However, it was suggested that on-farm research is needed into aspects of these practices to ensure that they create net sequestration of carbon; these related mainly to the relationship between soil carbon and nitrogen dynamics within the system. Although some of the scientific uncertainties relate to the specific environment and farming system, others are more general. In particular, it would be useful to have quicker and easier ways of measuring and estimating emissions of N_2O, both from fertilizer use and from deposition in and volatilization from urine and dung. It is also necessary to establish exactly which leguminous trees fix nitrogen, under what circumstances, and whether inoculation would be needed.

A theme that emerged from the study was the negative implications of dividing agricultural practices into conventional and agroecological approaches, of which experts might approve or disapprove. An additional but related theme has been that technology transfer needs to be understood as a process by which the transferred technology is the starting-point for the farmers' own innovation, so that it can be adapted to the specific farm. Rigid views on what the farmers should, or should not, do to qualify as carbon 'producers' will not assist that process.

The survey did reveal links between soil carbon and the broader economy. Labour shortages restricted what farmers would be able to try, especially at one of the survey locations. This reflects the way labour is no longer affordable on many farms. This may be linked not to rural–urban migration, but to poor commodity prices (in this case for milk) which have deprived farmers of the labour they need for sustainable management practices. This has in turn raised the question of who is really responsible for depleting the carbon in degraded pasture: the farmers, or the supermarkets paying too little for their milk? PES schemes for carbon sequestration might work in the region, but the carbon market may not be the only way to finance them; that might be done with a levy on the carbon's 'end user'.

The field survey has been useful in some respects, not least in identifying some underlying drivers for the exploitation of carbon stocks. But this book has considered many other aspects of carbon sinks in agriculture, and what is needed to make them work. They are reviewed in the final chapter.

Notes

1 Agrotoxins are understood here according to the official Brazilian definition: 'Products and agents of physical, chemical or biological processes, intended for use in production, storage and improvement of agricultural products, in pastures, in protection of forests (native or planted), and of other ecosystems of urban environments, hydrological or industrial, whose purpose is to alter the composition of flora or fauna with the objective of protecting them from damaging activity of living organisms considered noxious, as well as substances of products employed as defoliants, dessicants, and stimulators or inhibitors of growth' (quoted in Da Silva et al, 2005, p894).

2 The relevant document, *Instrução Normativa 07/99* of the Ministry of Agriculture, Livestock and Food Supply, is as follows: 'Agricultural and industrial production systems are considered organic in that they adopt technologies which optimize the use of natural and socioeconomic resources, respecting cultural integrity and having as an objective self-sufficiency in time and space, maximizing of social benefits, minimizing dependency on non-renewable energy and eliminating use of agrotoxins and other toxic artificial inputs, genetically-modified organisms (GMO/transgenics), or ionizing radiation in any phase of the process of production, storage and consumption, and at the same time prioritizing human and environmental health, assuring transparency in all stages of production and processing' (quoted in Feiden et al, 2002, p183). This is a normative definition and a farmer would not necessarily have to comply with all this to obtain certification.

3 *Ocupações Rural Não-Agrícolas*. This more accurately describes them than the English acronym used by Da Silva and Del Grossi (2001), RNFE (Rural Non-Farm Employment), as the latter does not make it clear if an off-farm occupation is divorced from agriculture altogether.

10

The Keys to Soil Carbon

No matter how complex any given situation may be, it should yield to analysis. We recognise that history cannot be reduced to an exact science of logic and law. Yet we hold that the causes and effects for most of the broad trends can be determined by the application of logic to the known facts. (Carter and Dale, *Topsoil and Civilization,* 1955)

This book has reviewed whether agriculture could be used, in theory, to mitigate climate change through the sequestration of carbon, and whether that would be a sensible model for the developing world. The short answer is yes to both.

This book has, however, identified significant constraints. One is common to all carbon trading; it is that, if the absorptive capacity of the atmosphere is no longer thought sufficient, then someone must decide how much can be allocated per head. This done, it will quickly acquire a market value. To do this, however, one must know how much of a commodity there is to trade, so someone must decide where emissions should be stabilized – say at 550ppm – and then work out how much carbon can be emitted without prejudicing this end. They must then decide who gets the permits to emit. In theory, the answer should be easy; if we can safely emit (say) 1.25 tonnes per capita, everyone gets that allowance, and must buy extra allowances on the market if they wish to emit more. This will not end the arguments about justice and equity; many believe that the richer countries should not be able to buy off their sins in this way, and perhaps they should not. But this seems to be the most effective way to ensure mitigation, and that is surely the priority.

Can agriculture be used for climate mitigation? It can, and much of this potential is in developing countries. Much more data are needed, but Lal (2004b, 2004c) has estimated soil carbon sequestration potential, including erosion control, in China at 126–364Tg/yr C, and that of India at 39.3–49.3Tg/yr C. A probably cautious estimate cited in Chapter 2 was that by Niles et al (2003, p79), who suggested that 420.6Mt could be sequestered during 2003–2013. This appeared to exclude large areas of grassland or rangeland in Central Asia and the Middle East.

These figures can be favourably compared with forestry projects. Net primary production (NPP) of biomass in cropland and grassland is comparable to that in tropical forests, and they surely provide a living for far more people than forests do. Indeed, if one wanted to be mischievous, one might suggest that the best way to retain forest carbon would be to cut down the trees. The cleared land could then be used for well-managed pasture, which in the long term would hold soil carbon at least as well as forest. Of course this argument would have flaws. Tropical forests are important biodiversity hotspots; besides, some people do rely on them for their livelihoods, and they may be from unique cultures that would not survive the forest's destruction. Nonetheless, sinks in cropland and grassland surely deserve more attention from policymakers than forests, not less.

This book has identified several key problems in the use of on-farm sinks for climate mitigation and agricultural development. The author believes, however, that they are soluble. This final chapter sums up the main challenges.

Financing the Carbon-Friendly Farm

The first difficulty, discussed in Chapter 5, is the high cost per tonne of mitigating climate change through agriculture. Updating a power station or funding alternative energy looks much better value, and will give much quicker returns; a carbon sink takes a long time to build. The returns are further reduced by the transaction costs of dealing with farmers, who may control enormous areas in (say) the United States or Ukraine, but are less likely to in most developing countries. Even in Brazil, where there are many very large units, they are not in the majority.

However, it seems most likely that such projects will be unilateral; that is to say, the project developer will proceed without investment from developed countries, and then certify the credits afterwards. They will then be sold on the open market at the price then prevailing for certified emissions. So the cost of mitigation per tonne will be immaterial. An analogy might be a tonne of coal; it is worth the same in the market once it has been extracted, whether this was done cheaply from a modern, efficient mine, or expensively, from a backward one. As the *State and Trends* reports from the World Bank have shown, carbon, once on the market, is increasingly treated this way. High mitigation costs per tonne are therefore no bar in theory to agricultural sinks. They are to the project developer, of course; but the assumption with agriculture is that for them, carbon will not be the only or even the main benefit. It will simply help finance projects that enhance sustainability and improve on-farm incomes.

There are two problems with this. One is that the project developer – perhaps not a group of farmers as such, more likely a national or regional Ministry of Agriculture – will have to find the finance, and in the case of any post-Kyoto mechanism it will probably not be allowed to come from overseas development assistance (ODA). This is because, under the Kyoto Protocol, the sustainable development benefits of the CDM are supposed to be additional to ODA, which is not meant to be diverted for this purpose. There is no reason to assume this will change after 2012. However, other trading mechanisms may be less concerned about this, and in any case, loans of types not considered to be ODA would presumably be eligible.

The second problem with the unilateral model is it involves the simple production of a commodity by the host country – in this case, carbon credits – and their sale on a market where they are fungible. One might argue that this is selling a primary commodity instead of exploiting it oneself and that one therefore surrenders part of its value for no return. This would probably be the view of a dependency theorist such as Raúl Prebisch or Andre Gunder Frank. Even if one does not place the transaction within a theoretical framework in that way, a unilateral project would not directly provide any of the sustainable development benefits through technology transfer that foreign investment in Kyoto-based projects was supposed to provide. As mentioned in Chapter 5, some observers in India have questioned this, given the large percentage of CDM projects there that are financed in this way.

But perhaps this was always going to be. If policymakers enlist the market to help them reach their goals, they may have to accept its inherent logic; it is not, in its most meaningful form, something that governments control. If one creates a product that is fungible, one cannot always dictate the way it is bought and sold. In short, the transaction costs and high costs per tonne of agricultural projects need not keep them out of the market; they will just make less of a profit per tonne. The fact that this partly subverts sustainable development objectives will have to be accepted.

Scientific Uncertainty and the Challenge of Heterogeneity

A significant challenge is that the uncertainty that surrounds climate change extends from the fundamental principles of climate science right down to the carbon sequestration potential of individual agronomic practices. There is inadequate knowledge of some processes that greatly affect the relationship between land use and climate, in particular soil erosion and nitrogen cycling. There can be no business in which the manufacturing plant and processes are as heterogeneous as agriculture. Uncertainty arises in several key mitigation areas.

First of all, as discussed in Chapter 2, there is serious disagreement as to whether soil erosion is a net source or sink. This may be moot; what matters is the net flux in the particular watershed in which a mitigation activity takes place. But it could also be argued that the global effect on the terrestrial carbon cycle needs to be known for policy purposes; how much effort should we devote to this? Even if one dismisses this and accepts that the net flux need only be known at watershed level, there are still serious challenges both in measuring the extent of soil erosion and in quantifying the fate of the translocated SOC. These were reviewed in Chapter 5. There is doubt in some quarters as to whether soil erosion can be accurately measured at all. The author does not go that far, but does accept that the methodological challenges are severe.

Second, further uncertainties were discussed in the subsections of Chapter 8 that dealt with inorganic fertilizer and pasture improvement. These uncertainties concerned the net sequestration possible from improving pasture, with or without chemical fertilizer. But this may increase the carrying capacity of the pasture. The effect of pasture improvement on net fluxes of CO_2e are governed in part by stocking rates; these can of course increase net emissions by degrading the pasture and by raising enteric CH_4 emissions. But is this always the case? The chapter quoted evidence that heavier grazing can *reduce* the presence of invasive species, to the advantage of grasses that sequester more CO_2 as SOC.

At the same time, however, the quantitative relationship between pasture improvement, stocking rates and net GHG fluxes is governed by a series of complex interactions between carbon sequestration, enteric CH_4 emissions and volatilization of N from dung and urine – and the latter factor may be further complicated by the extent to which cattle foregather for feed and water, trampling the ground in a given location and affecting its uptake capacity. In this situation, who would guarantee to accurately quantify net GHG fluxes from a hectare of pasture? To make matters worse, there is the question of whether intensification of pasture at the forest frontier increases or reduces deforestation – a question that was reviewed briefly in Chapter 5, and has profound implications for the successful implementation of carbon payments for avoided deforestation.

Third, and potentially very contentious, are the questions surrounding tillage regimes. These are bundled together, in this book, under the title of conservation tillage (CT) but may be referred to as conservation agriculture, reduced tillage or no-till; the name used can affect the exact package of practices referred to, and the rigidity with which that package is defined. But all refer to reduction or elimination of tillage, together with retention of crop residues. CT regimes are often advocated both for soil conservation in general, and for carbon retention and sequestration. However, as explained in Chapter 8, there are some doubts as to the exact circumstances under which CT does build up SOC, where in the soil horizon it will accumulate (which could affect permanence), and the mechanisms under which it may or may not move between those horizons. Besides these biophysical questions, there are also farming-systems complications, particularly around the retention of crop residues in place of their use for feed; this could have social implications and could also cause carbon leakage. Further such questions arise from the need for rotations.

A fourth scientific uncertainty, discussed in Chapter 2, concerns the potential of biochar. Before this becomes a widespread mitigation strategy in agriculture, there will be a need for greater understanding of how its use creates *terra preta* or *terra mulata*, how to ensure that biochar's use does not result in unhelpful biological changes in the soil, and how to prevent the introduction of toxins. Methodologies will also be needed for calculating the net carbon fluxes from the production and addition of biochar.

These uncertainties have two main consequences for carbon sinks in agriculture. The first is that carbon sequestration using any of these technologies is likely to be highly site-specific. CT is a case in point. One farmer implements it with one package of rotations, his neighbour adopts it but with a different package, and a third adopts a mixture of the two but decides that he will not or cannot stop pastoralists or other farmers from grazing his barley stubble, so does not leave the crop residues in the field. There will clearly be different sequestration rates per hectare.

It therefore becomes difficult to base payments on adoption of a given technology with a notional accumulation of carbon. That will not satisfy the demanding methodologies for certified emissions reductions. The reductions produced, therefore, may or may not be certified for compliance purposes. If they can't be, they will not have a high market value. They *can* be sold, for less money, on the voluntary non-compliance-based market; a good enterprise – but one that will never leverage the vast sums earned by compliance credits through the EU ETS.

The implication is that, for such markets, projects will have to actually measure carbon accumulation against the business-as-usual (BAU) baseline. As discussed in Chapter 5, this is theoretically easy but practically challenging, not least due to the spatial variation found in even quite homogeneous environments. Does this mean the use of agricultural carbon in commercial compliance markets is impractical?

The author does not believe so, for two basic reasons. One is the inevitability of uncertainty at the frontiers of any science or technology. This may be seen in the arguments about climate change itself; the uncertainty of climate science is sometimes highlighted by the sceptics – but uncertainty is not a good argument for scepticism about climate change, or for not mitigating it. Those who regard anthropogenic climate change as an unproven thesis on the basis that causality is not explicit, are misunderstanding a basic reality of research on the cutting edge: that it *must* proceed on a hypothesis and be inductive. As Bertrand Russell pointed out, we cannot generate new knowledge in any other way; indeed, he argues that 'all the important inferences outside logic and pure mathematics are inductive, not deductive'; the only exceptions are, he says, law and theology, 'each of which derives its first principles from an unquestionable text' (Russell, 1946, p222). What applies to climate change itself, also applies to the measures we might use to mitigate it. If it is not known just how CT sequesters carbon, the answer is not to shrug one's shoulders. The answer is to find out. This spirit has also been expressed by Kimble in his contention, quoted in Chapter 5, that soil scientists must explain the causes of variability in soil samples 'and not use it as a crutch for not wanting to answer policy questions' (Kimble, 2006, p488).

In any case, carbon sequestration is a biophysical process that can surely be measured. Albrecht's 1938 description of soil processes, quoted in Chapter 2, implied that we may view a farmed field as a factory, producing food out of carbon, and that this process is quantifiable. This is not conceptually different from measuring the inputs and outputs from a power station. Moreover, as stated above, the process of carbon accumulation in soil can be measured in principle. When a researcher such as Lal (2007, p303) puts global mitigation potential as high as 0.6–1.2Pg C per year, there is great uncertainty in such estimates – but they are not speculation. Moreover sequestration of carbon should be considered along with reduction of emissions from soil erosion, despite the enormous uncertainties and difficulties that this presents.

Having concluded that agricultural sinks should be measured, this does not imply that other methods – based around adoption of an approved practice – have no place. They may not have access to the large sums generated by compliance markets (though that is not certain; methodologies may yet be developed). However, they have several advantages. First, although there will still be monitoring and verification demands, they will not be so severe. Second, because they are not priced on the basis of sequestered carbon, they can factor-in other environmental goods that are harder to price. These include agrobiodiversity; this has concrete financial value, but it is hard to internalize in a transaction. Reductions in agrotoxins and protection of urban water supplies are also important goods. Third, schemes of this sort have a role in REDD+, in which payments for preservation of the forest can be combined with subsidies to agriculture. Last but not least, this type of activity can successfully be pursued using public funding, either international (for example through GEF), or from national governments. In the latter case, government might make these sort of payments for their own purposes (as local authorities in Mexico and the United States have for protection of water supplies), or they could disburse REDD+ funding.

However, all such programmes will have a disadvantage – they are predicated on getting farmers to adopt certain predefined practices, or packages of them. As stated above, this makes assumptions about sequestration rates that may sometimes not be warranted. Of course, if PES schemes are not selling carbon by the exact tonne, this may not matter much. What does, is the way in which different practices may fit into one farmer's farming system and be quite inappropriate for their neighbour, even when the two farms seem quite similar. Hidden differences – small variations in sloping land, or, especially, availability of labour due to larger or smaller family size – mean that the farmer will have to be able to pick and choose between technologies. The field survey described in

Chapters 6–9 seemed to confirm this. Indeed, as we have seen, the farmer may need to be able to discriminate between *components* of technologies, although their enthusiasts may prefer to promote them as packages; a point also well made with regard to CT by Giller et al (2009). Land users should be able to tailor technologies to their individual requirements.

It might also be helpful if decision-makers accepted their ability to do so. As recorded in Chapter 7, some staff in the survey area in Brazil were not complimentary about farmers' willingness to innovate. But, when asked, farmers *did* show interest in practices they had not encountered before, such as the leguminous trees. Of course they did not endorse them as such, but that would be agreeing instantly to modify their most productive assets. This point would be unsurprising if made in relation to any other type of businesses – the motor industry, for example. It may be that experts should stop viewing the farm as different from any other enterprise that needs to make decisions about investment in capital equipment in the light of its income and outgoings.

This, then, is one of the major problems for agricultural sinks identified in this book; the challenges of scientific uncertainty and of heterogeneity, and the way in which they are interrelated. Biophysical research can help illuminate sources of variability in carbon sequestration and can also improve monitoring and verification methodologies. It should be a high priority.

Some Broader Issues

The science question may lead into discussion of some of the underlying issues around climate change and development that arise from this book.

There are schools of thought that undervalue physical science. This is especially so with soil, which can generate a certain mysticism. There is nothing new in this. In her account of her eccentric upbringing, *Hons and Rebels*, Jessica Mitford quotes her rather strange Uncle Geoffrey, in whose view national character was closely related to the use of manure: 'It is an actual fact that character is largely a product of the soil. ... Chemicals have had their poisonous day. It is now the worm's turn to re-form the manhood of England,' he wrote (Mitford, 1960, p19). This mysticism is also suggested by Lowdermilk (1948), who travelled through the Middle East for the US government in 1939. Giving a radio talk on soil conservation in Jerusalem in June of that year, he gave what he called the Eleventh Commandment: 'Thou shalt … safeguard thy fields from soil erosion … If any shall fail … thy fruitful fields shall become sterile stony ground and wasting gullies, and thy descendants … live in poverty or perish from off the face of the earth.'

This evangelical mysticism now finds its expression in radical environmentalism. For example, M. Lewis (1992, pp28–29) attacks the school of thought known as deep ecology; this, briefly stated, is the view that humanity is part of 'an interconnected totality, the whole of which is much greater than the sum of its parts'. 'While celebrating diversity and toleration,' says Lewis, 'deep ecologists dismiss virtually the entire heritage of Western thought as morally bankrupt and leading us to ecological destruction' (M. Lewis, 1992, pp28–29). He later adds (M. Lewis, 1992, p135) that the 'eco-radical' attack on reductionist, specialist approaches to science is a threat in itself because it opposes the scientific methodology that can alone answer the challenge of the environment.

The author cannot find anything wrong with seeing life on earth as interconnected phenomena. Neither is there any reason why structures of thought should not be open to question; in any case, deep ecology is an extreme example. Nonetheless, Lewis has a point; there is a touch of the anti-science in many quite moderate mainstream movements – not least among some proponents of organic agriculture.

Yet stress on biophysical science is necessary. If it has had its failures, that is not because it is wrong in principle, but because someone has missed the need to connect different disciplines and fields of enquiry. This was clearly the case with the failed soil-conservation projects described in Chapter 7. But one can accept that environmental science should be multidisciplinary – which is the author's own view – without underestimating the need for specialization in Western science. Few people can acquire expertise in both (say) estimating N_2O emissions from fertilizer, and the measurement of soil erosion. As Vandermeer (1995, p216) has put it, 'it is important to distinguish between reductionism as a research tactic and reductionism as a philosophy – the latter is the folly of modernism, the former is a necessity of science'. Many of the gaps in knowledge that affect agricultural sinks, for example on the relationship between soil erosion and emissions, or biological nitrogen fixation by leguminous trees, do demand specialization – and reductionism. The widespread application of biophysical as well as social science is the most urgent single requirement for mitigating climate change.

However, regarding climate mitigation, agriculture is an interesting case study in another respect too – the fact that climate negotiators do not really appear to understand the issues. This extends right up to the highest level. At present, even where countries have mitigation commitments, there is no automatic incentive to preserve sinks in non-forest land, as national reporting of Article 3.4 (e.g. terrestrial but non-forest) sinks is not compulsory. True, if a country *does* include such sinks in its national reporting, it must continue to do so thereafter. However, governments may still manipulate the system, as they can report CO_2 removals but not de-vegetation that causes emissions (Cowie et al, 2007b, pp307, 309; Höhne et al, 2007, p354). This means that credits may be obtained for a 10ha plot while the surrounding 100ha deteriorates. This is a strange logic. As Schlamadinger et al (2007b, p280) put it: 'If the aim of a climate treaty is to protect the global atmosphere for future generations, it is important that accounting be comprehensive over time and space.' It is an indication of the dangerous confusion around land-use issues under the Kyoto Protocol, and the need to tidy it up.

This has a further implication – that the climate negotiating process might not be able to cope with the complexities it has to deal with. Britain's former ambassador to the United States, Sir Christopher Meyer, has described how nation speaks unto nation. 'Multilateral diplomacy ... has its moments,' he says. 'But the price is high: listening to droning diplomatic windbags; crawling line-by-line through interminable texts; an obsession with procedural points; rooms stuffy with bad air and the ineffable smugness of a cosy club; and the painful difficulty of reaching a common decision' (Meyer, 2007, p46).

In this atmosphere, important detail will get lost. It is a truism that climate science is complicated, but this book has shown that this extends right down to unravelling such matters as the relationship between fertilizer, soil type and a few extra cattle in a given pasture. It is therefore perhaps not surprising that the negotiators of the Marrakech Accords shied away from Article 3.4 activities (although this was also due to the vagaries of the negotiating process, as explained in Chapter 4). But it was unfortunate, as an important opportunity for mitigation was missed.

Today, with the expiry of the original Kyoto Protocol very close, there is a chance to reconsider. This time, it should not be allowed to slip away.

References

Adams, W. (2001) *Green Development* (2nd edn), Routledge, London

Adger, W., Brown, K., Fairbrass, J., Jordan, A., Paavola, J., Rosendo, S. and Seyfang, G. (2002) *Governance for Sustainability: Towards a 'Thick' Understanding of Environmental Decision-Making*, CSERGE Working Paper EDM 0204, Centre for Social and Economic Research on the Global Environment, University of East Anglia, Norwich

Agarwal, A. and Narain, S. (1991) *Global Warming in an Unequal World: A Case of Environmental Colonialism*, Centre for Science and the Environment, New Delhi, India

Albrecht, W. (1938) *Soils and Men, Yearbook of Agriculture 1938*, United States Department of Agriculture, Washington, USA

Altieri, M. (2002) 'Agroecology: The science of natural resource management for poor farmers in marginal environments', *Agriculture, Ecosystems and Environment*, vol 93, pp1–24

Angelsen, A. (2009) 'Policy options to reduce deforestation', in A. Angelsen (ed)*Realising REDD+: National Strategy and Policy Options*, CIFOR, Bogor, Indonesia, pp125–138

Araujo Castro, J. (1972/1998) 'Environment and development: The case of the developing countries', in K. Conca and G. Dabelko (eds) *Green Planet Blues: Environmental Politics from Stockholm to Kyoto*, Westview Press, Boulder, USA, pp32–39

Armenteras, D., Gast, F. and Villareal, H. (2003) 'Andean forest fragmentation and the representativeness of protected natural areas in the eastern Andes, Colombia', *Biological Conservation*, vol 113, pp245–256

Armando, M. (2002) *Ferramenta para uma Agricultura Sustentável*, EMBRAPA Brasília, Brazil (in Portuguese)

Arndt, D., Baringer, M. and Johnson, M. (eds) (2010) 'State of the climate in 2009', *Bulletin of the American Meterological Society*, vol 91, no 7, S1–S224

Arrhenius, S. (1903) *Development of the Theory of Electrolytic Dissociation*, Nobel lecture, delivered on 11 December 1903. Downloaded on 1 December 2005 from http://nobelprize.org/chemistry/laureates/1903/arrhenius-lecture.pdf

Atjay, G., Ketner, P. and Duvigneaud, P. (1979) 'Terrestrial primary production and phytomass', in B. Bolin, E. Degens, S. Kempe and P. Ketner (eds.) *The Global Carbon Cycle*, John Wiley & Sons, Chichester, pp129–181.

Azar, C. and Schneider, S. (2003) 'Are the economic costs of (non-)stabilising the atmosphere prohibitive? A response to Gerlagh and Papyrakis', *Ecological Economics*, vol 46, pp329–332

Baker, J., Ochsner, T. and Venterea, R. (2007) 'Tillage and soil carbon sequestration – What do we really know?', *Agriculture, Ecosystems and Environment*, vol 118, pp1–5

Baliunas, S. (2004) *Witches and Weather*, Address delivered at the conference 'Risk: Regulation and Reality', Toronto, 7 October, 2004 www.ideasinactiontv.com/tcs_daily/2005/01/witches-and-weather.html, accessed 10 March 2010

Bationo, A., Kihara, J., Vanlauwe, B., Waswa, B. and Kimetu, J. (2007) 'Soil organic carbon dynamics, functions and management in West African agro-ecosystems', *Agricultural Systems*, vol 94, pp13–25

Batjes, N. (2001) 'Options for increasing soil carbon sequestration in West African soils: An exploratory study with special focus on Senegal', *Land Degradation & Development*, vol 12, pp131–142

Baumert, K. and Pershing, J. (2004) *Climate Data: Insights and Observations*, Pew Center, Washington, DC, USA

Baumert, K., Kete, N. and Figueres, C. (2000) *Designing the Clean Development Mechanism to Meet the Needs of a Broad Range of Interests*, World Resources Institute Climate Notes, Pew Center, Washington, DC, USA

Berhe, A., Harte, J., Harden, J. and Torn, M. (2007) 'The significance of the erosion-induced terrestrial carbon sink', *BioScience*, vol 57, no 4, pp337–346

Berk, M. and den Elzen, M. (2001) 'Options for differentiation of future commitments in climate policy: How to realise timely participation to meet stringent climate goals?' *Climate Policy*, vol 1, pp465–480

Berthelot, M., Lingstein, P., Ciais, P., Dufresne, J. and Monfray, P. (2005) 'How uncertainties in future climate predictions translate into future terrestrial carbon fluxes', *Global Change Biology*, vol 11, pp959–970

Bewket, W. and Sterk, G. (2002) 'Farmers' participation in soil and water conservation activities in the Chemoga watershed, Blue Nile Basin, Ethiopia', *Land Degradation & Development*, vol 13, no 3, pp189–200

Bhatt, R., Baig, M. and Tiwari, H. (2007) 'Growth, biomass production, and assimilatory characters in *Cenchrus ciliaris* L. under elevated CO_2 conditions', *Photosynthetica*, vol 45, pp296–298

Blaikie, P. (1985) *The Political Economy of Soil Erosion in Developing Countries*, Longman, Harlow

Blaikie, P. (1989) 'Explanation and policy in land degradation and rehabilitation for developing countries', *Land Degradation & Rehabilitation*, vol 1, no 1, pp23–38

Bloomfield, J. and Pearson, H. (2000) 'Land use, land-use change, forestry, and agricultural activities in the Clean Development Mechanism: Estimates of greenhouse gas offset potential', *Mitigation and Adaptation Strategies for Global Change*, vol 5, pp9–24

BNDES (2010) 'Programa para Redução da Emissão de Gases de Efeito Estufa na Agricultura – Programa ABC', www.bndes.gov.br/SiteBNDES/bndes/bndes_pt/Institucional/Apoio_Financeiro/Programas_e_Fundos/abc.html (in Portuguese), accessed 28 December 2010

Boddey, R., Sá, J., Alves, B. and Urquiaga, S. (1997) 'The contribution of biological nitrogen fixation for sustainable agricultural systems in the tropics', *Soil Biology and Biochemistry*, vol 29, no 5/6, pp787–799

Boddey, R., Xavier, D., Alves, B. and Urquiaga, S. (2003) 'Brazilian agriculture: The transition to sustainability', *Journal of Crop Production*, vol 9, pp593–621

Boddey, R., Macedo, R., Tarré, R., Ferreira, E., de Oliveira, O., Rezende, P., Cantarutti, R., Pereira, J., Alves, B. and Urquiaga, S. (2004) 'Nitrogen cycling in *Brachiaria* pastures: The key to understanding the process of pasture decline', *Agriculture, Ecosystems and Environment*, vol 103, pp389–403

Boddey, R., Jantalia, C., Macedo, M., de Oliveira, O., Resende, A., Alves, B. and Urquiaga, S. (2006) 'Potential for carbon sequestration in soils of the Atlantic Forest region of Brazil', in R. Lal, C. Cerri, M. Bernoux, J. Etchevers and E. Cerri (eds) *Carbon Sequestration in Soils of Latin America*, Food Products Press, Binghampton, NY, USA, pp305–348

Boyd, E. (2002) 'The Noel Kempff project in Bolivia: Gender, power, and decision-making in climate mitigation', *Gender and Development*, vol 10, no 2, pp70–77

Brick, S. (2010) *Biochar: Assessing the Promise and Risks to Guide U.S. Policy*, Natural Resources Defense Council, Washington, DC, USA

Brown, K. (1997) 'The road from Rio', *Journal of International Development*, vol 9, no3, pp383–389

Brown, K. and Corbera, E. (2003) *A Multi-Criteria Assessment Framework for Carbon-Mitigation Projects: Putting 'Development' in the Centre of Decision-Making*, Tyndall Centre Working Paper 29, The Tyndall Centre, Norwich, UK

Brown, K., Tompkins, E. and Adger, N. (2002) *Making Waves: Integrating Coastal Conservation and Development*, Earthscan, London, UK

Brown, K., Adger, W. N., Boyd, E., Corbera, E. and Shackley, S. (2004) *Evaluating Policy Options for the Clean Development Mechanism: A Stakeholder Multi-Criteria approach*, Technical Report 16, The Tyndall Centre, Norwich, UK

Bryson, B. (1998) *Notes from a Big Country*, Transworld Publishers, London, UK

Buainan, A., Romeiro, A. and Guanzirou, C. (2003) 'Agricultura familiar e o novomundo rural', *Sociologias*, vol 5, no10, pp312–347 (In Portuguese)

Buchholz, T. (1999) *New Ideas from Dead Economists*, Penguin, London, UK

Burleson, E. (2008) 'The Bali climate change conference', *American Society of International Law Insights*, vol 12, no 4, 17 March 2008

Butzengeiger, S. (2005) *Voluntary Compensation of GHG Emissions: Selection Criteria and Implications for the International Climate Policy System*, Hamburg Institute of International Economics, Hamburg, Germany

Câmara, I. (2003) 'Brief history of conservation in the Atlantic Forest', in C. Galindo-Leal and I. Câmara (eds) *The Atlantic Forest of South America – Biodiversity Status, Threats, and Outlook*, Island Press, Washington, DC, USA, pp31–42

Campbell, J. (1979) 'Spatial variability of soils', *Annals of the Association of American Geographers*, vol 69, no 4, pp544–556

Canadell, J., Kirschbaum, M., Kurz, W., Sanz, M-J., Schlamadinger, B. and Yamagata, Y. (2007) 'Factoring out natural and indirect human effects on terrestrial carbon sources and sinks', *Environmental Science & Policy*, vol 10, pp370–384

Capoor, K. and Ambrosi, P. (2006a) *State and Trends of the Carbon Market 2006*, The World Bank, Washington, DC, USA

Capoor, K. and Ambrosi, P. (2006b) *State and Trends of the Carbon Market 2006 (Update: January 1–September 30 2006)*, The World Bank, Washington, DC, USA

Capoor, K. and Ambrosi, P. (2007) *State and Trends of the Carbon Market 2007*, The World Bank, Washington, DC, USA

Carbon Dioxide Information Analysis Center (2003) 'Frequently asked questions', Carbon Dioxide Information Analysis Center, http://cdiac.esd.ornl.gov/ pns/faq.html, accessed 25 March 2004

Carmona, R. and Zatz, R. (1998) 'Sistemas de preparo do solo e controle de plantas daninhas perenes em *Brachiaria decumbens', Pesquisa Agropecuária Brasileira*, vol 33, no 9, pp1515–1523 (in Portuguese)

Carpenter, R. (1989) 'Do we know what we are talking about?', *Land Degradation & Rehabilitation*, vol 1, no 1, pp1–4

Carpentier, C. L., Valentim, J., Vosti, S. A. and Witcover, J. (2000) 'Intensified small-scale cattle production systems in the western Brazilian Amazon: Will they be adopted and can they save the forest?', *Agriculture, Ecosysystrems and Environment*, vol 82, pp73–88

Carr, M., Brash, K. and Anderson, R. (2010) *Climate Change: Addressing the Major Skeptic Arguments*, DB Climate Change Advisors, Frankfurt am Main, Germany

Carter, V. and Dale, T. (1955) *Topsoil and Civilization* (revised edn, 1985), University of Oklahoma Press, Norman, OK, USA

Carvalho, P. (2006) 'Brazil', FAO Country Pasture/Forage Resource Profiles, www.fao.org/ag/AGP/AGPC/ doc/Counprof/Brazil/brazil.htm#1.int, accessed 12 April 2007

Cassel, A. and Patel, R. (2003) *Agricultural Trade Liberalization and Brazil's Rural Poor: Consolidating Inequality*, Policy Brief No. 8, Institute for Food and Development Policy, Oakland, USA

Chan, K. and Xu, Z. (2009) 'Biochar: Nutrient properties and their enhancement', in J. Lehmann and S. Joseph (eds) *Biochar for Environmental Management*, Earthscan, London, pp67–84

Chandler, W., Schaeffer, R., Dadi, Z., Shukla, P., Tudela, F., Davidson, O. and Alpan-Atamer, S. (2002) *Climate Change Mitigation in Developing Countries: Brazil, China, India, Mexico, South Africa, and Turkey*, Pew Center for Global Climate Change, Washington DC, USA

Cheng, H. and Kimble, J. (2001) 'Characterization of soil organic carbon pools', in R. Lal, J. Kimble, R. Follett and B. Stewart (eds) *Assessment Methods for Soil Carbon*, CRC Press, Boca Raton, USA, pp117–130

Chomitz, K., Buys, P., De Luca, G., Thomas, T. and Wertz-Kanounnikoff, S. (2007) *At Loggerheads? Agricultural Expansion, Poverty Reduction, and Environment in the Tropical Forests*, The World Bank, Washington, DC, USA

CO_2 Science (2007) *The Most Important Fodder Crop of the Arid and Semi-arid Tropics*, downloaded on September 4 2007 from www.co2science.org/scripts/ CO2Science B2C/articles/V10/N32/B1.jsp

Conant, R. and Paustian, K. (2002) 'Spatial variability of soil organic carbon in grasslands: Implications for detecting change at different scales', *Environmental Pollution*, vol 116, S127–S135

Corbera, E. (2004) 'Bringing development into carbon forestry markets: Challenges and outcomes of small-scale carbon forestry activities in Mexico', in D. Murdiyarso and

H. Herawati (eds) *Carbon Forestry: Who Will Benefit?*, CIFOR Bogor, Indonesia, pp42–56

Corfee-Morlot, J. and Höhne, N. (2003) 'Climate change: Long-term targets and short-term commitments', *Global Environmental Change*, vol 13, pp277–293

Cóser, A., Cruz Filho, A., Martins, C., Carvalho, L., Alvim, M. and Freitas, V. (1997) 'Desempenho animal em pastagens de capim-gordura e braquiária', *Pasturas Tropicales,* vol 19, no 3, pp14–19

Costa, F., Gomes, J., Bayer, C. and Mielniczuk, J. (2006) 'Métodos para avaliação das emissões de gases do efeito estufa no sistema solo-atmosfera', *Ciência Rural*, vol 36, no 2, pp693–700 (in Portuguese)

Costanza, R., Cumberland, J., Daly, H., Goodland, R. and Norgaard, R. (1997a) *An Introduction to Ecological Economics*, St Lucie Press, Boca Raton, USA

Costanza, R., d'Arge, R., de Groot, R., Farber, S., Grasso, M., Hannon, B., Limburg, K., Naeem, S., O'Neill, R., Paruelo, J., Raskin, R., Sutton, P. and van den Belt, M. (1997b) 'The value of the world's ecosystem services and natural capital', *Nature*, vol 387, pp253–260

Cowie, A., Schneider, U. and Montanarella, L. (2007a) 'Potential synergies between existing multilateral environmental agreements in the implementation of land use, land-use change and forestry activities', *Environmental Science & Policy*, vol 10, pp335–352

Cowie, A., Kirschbaum, A. and Ward, M. (2007b) 'Options for including all lands in a future greenhouse gas accounting framework', *Environmental Science & Policy*, vol 10, pp306–321

Critchley, W. (2000) *Groundtruthing: New Perspectives on Soil Erosion and Conservation in the Tropics*, Vrije Universiteit te Amsterdam, Amsterdam, The Netherlands

Critchley, W. and Mutunga, K. (2003) 'Local innovation in a global context: Documenting farmer initiatives in land husbandry through WOCAT', *Land Degradation & Development*, vol 14, pp143–163

Crosson, P. and Anderson, J. (2002) *Technologies for Meeting Future Global Demands for Food*, Discussion Paper 02-02 of the International Food Policy Research Institute, Washington, DC, USA

Da Silva, J. and Del Grossi, M. (2001) 'Rural nonfarm employment and incomes in Brazil: Patterns and evolution', *World Development*, vol 29, no 3, pp443–453

Da Silva, J., Del Grossi, M. and Campanhola, C. (2002) 'O que há realmente novo no rural Brasileiro', *Cadernos de Ciência & Tecnologia (Brasília)*, vol 19, no 1, pp37–67 (in Portuguese)

Da Silva, J., Novato-Silva, E., Faria, H. and Pinheiro, T. (2005) 'Agrotóxico e trabalho: Uma combinação perigosa para a saúde do trabalhador rural', *Revista Ciência e Saúde Coletiva*, vol 10, no 4, pp891–903 (in Portuguese)

De Alcântara, F., Neto, A., De Paula, M., De Mesquita, H. and Muniz, J. (2000) 'Adubação verde na recuperação da fertilidade de um latossolo vermelho-escuro degradado', *Pesquisa Agropecuária Brasileira*, vol 35, no 2, pp277–288 (in Portuguese)

De Graaf, J. (1999) 'Evaluating incentive systems for soil and water conservation on the basis of case studies in four countries', in D. Sanders, P. Huszar,

S. Sombatpanit and T. Enters (eds) *Incentives in Soil Conservation: From Theory to Practice*, Science Publishers, Enfield, USA, pp101–118

Dean, W. (1997) *With Broadax and Firebrand: The Destruction of the Brazilian Atlantic Forest*, University of California Press, Berkeley, USA

Dessler, A. and Parson, E. (2006) *The Science and Politics of Global Climate Change: A Guide to the Debate*, Cambridge University Press, Cambridge, UK

Díaz-Zorita, M., Duarte, G. and Grove, J. (2002) 'A review of no-till systems and soil management for sustainable crop production in the subhumid and semiarid Pampas of Argentina', *Soil & Tillage Research*, vol 65, pp1–18

Dlugokencky. E. (2010) 'Carbon dioxide, methane, and carbon monoxide', in D. Arndt, M. Baringer and M. Johnson (eds) *State of the Climate in 2009*, Special Supplement to the Bulletin of the American Meteorological Society, vol 91, pp41–44

Doherty, R., Sitch, S., Smith, B., Lewis, S. and Thornton, P. (2010) 'Implications of future climate and atmospheric CO_2 content for regional biogeochemistry, biogeography and ecosystem services across East Africa', *Global Change Biology*, vol 16, no 2, pp617–640

Doraswaimy, P., McCarty, G., Hunt, E., Yost, R., Doumbia, M. and Franzluebbers, A. (2007) 'Modeling soil carbon sequestration in agricultural lands of Mali', *Agricultural Systems*, vol 94, pp63–74

Downie, A., Crosky, A. and Munroe, P. (2009) 'Physical properties of biochar', in J. Lehmann and S Joseph (eds) *Biochar for Environmental Management*, Earthscan, London, UK, pp13–29

Dubois, J. and Lamego, R. (1998) *Desenvolvimento Sustentável em Regiões Serranas do Rio de Janeiro: Aspectos Econômicos, Socioculturais e Políticas Oficiais de Uso da Terra*, Paper presented at Sathla Conference, Rio de Janeiro, 9–13 March, 1998 (in Portuguese)

Dubrovsky, M., Zalud, Z., Stasna, M. and Trnka, M. (2000) 'Effect of climate change and climate variability on crop yields', in *Proceedings of the 3rd European Conference on Applied Climatology*, Pisa, Italy, 16–20 October 2000

Dudley, R. (2005) *A Generic Look at Payments for Ecosystem Services: Plan or Scam?* Paper prepared for presentation at the 23rd International Systems Dynamics Conference, 17–20 July 2005, Boston, United States

Dutschke, M. (2002) 'Fractions of permanence – squaring the cycle of sink carbon accounting', *Mitigation and Adaptation Strategies for Global Change*, vol 7 pp381–402

Earth Observatory (2005a) 'Svante Arrhenius', http://earthobservatory.nasa.gov/Library/Giants/Arrhenius, accessed 3 December 2010

Earth Observatory (2005b) 'Roger Revelle', http://earthobservatory.nasa.gov/Library/Giants/Revelle, accessed 3 December 2010

Ebinger, M., Cremers, D., Meyer, C. and Harris, R. (2006) 'Laser-induced breakdown spectroscopy and applications for soil carbon measurement', in R. Lal, C. Cerri, M. Bernoux, J. Etchevers, and E. Cerri (eds) *Carbon Sequestration in Soils of Latin America*, Food Products Press, Binghampton, NY, USA, pp407–422

Edmeades, D. (2003) 'The long-term effects of manures and fertilizers on soil productivity and quality: A review', *Nutrient Cycling in Agroecosystems*, vol 66, pp165–180

Elings, A., Witcombe, J. and Stam, P. (2000) 'Use of genetic diversity in crop improvement', in C. Almekinders and W. De Boef (eds) *Encouraging Diversity*, Intermediate Technology Publications, Rugby, UK, pp20–25

Elliott, C. (1996) 'Paradigms of forest conservation', *Unasylva*, vol 47, no 4, pp3–9

EMBRAPA (2003) 'Cultivo de Tomate para Industrialização', http://sistemasdeproducao.cnptia.embrapa.br/FontesHTML/Tomate/Tomate Industrial/plantio.htm (in Portuguese), accessed 6 June 2007

Enters, T. (1999) 'Incentives as policy instruments – key concepts and definitions', in D. Sanders, P. Huszar, S. Sombatpanit and T. Enters (eds) *Incentives in Soil Conservation: From Theory to Practice*, Science Publishers, Enfield, USA, pp25–40

Ernsting, A. and Smolker, R. (2009) *Biochar for Climate Change Mitigation: Fact or Fiction?* Biofuelwatch, UK/USA

Esteves, B. (2007) 'Zero tillage: Brazil's own green revolution', SciDev.net, www.scidev.net/features /index.cfm? fuseaction=printarticle&itemid =576&language=1, accessed 6 June 2007

European Commission (2006) 'Greenhouse Gas Emission Allowance Trading Scheme', Directive 2003/87/EC of the European Parliament and of the Council of 13 October 2003, establishing a scheme for greenhouse gas emission allowance trading within the Community, http://europa.eu/legislation_summaries/energy/european_energy_policy/l28012_en.htm, accessed 12 December 2010

Evangelista, A., de Lima, J. and de Lavras, F. (2001) *Recuperação de pastagens degradadas*, Universidade Federal de Lavras, Lavras, Brazil (in Portuguese)

Evans, C. (2004) *Productive Forest Corridors in the Atlantic Forest, Brazil*, Iracambi Atlantic Rainforest Research and Conservation Center, Rosário da Limeira, Brazil

Evers, G. and Agostini, A. (2001) 'No-tillage farming for sustainable land management: Lessons from the 2000 Brazil study tour', FAO, Rome, Italy

FAO (1994) *Land Degradation in South Asia: Its Severity, Causes and Effects Upon the People*, World Soil Resources Report no 78, FAO, Rome, Italy

FAO (1998) 'FAO: Conventional tilling severely erodes the soil; New concepts for soil conservation required', www.fao.org/WAICENT/OIS/PRESS_NE/PRESSENG/1998/pren9842.htm, accessed 24 January 2011

FAO (2001a) *State of the World's Forests 2001*, FAO, Rome, Italy

FAO (2001b) 'Saving the Earth', FAO Newsroom Historic Archives, www.fao.org/english/newsroom/highlights/2001/011103-e.htm, accessed 23 January 2011

FAO (2002a) 'Hold back the desert with conservation agriculture', www.fao.org/english/newsroom/news/2002/10502-en.html, accessed 23 January 2011

FAO (2002b) *World Agriculture: Towards 2015/2030*, FAO, Rome, Italy

FAO (2002c) *The State of Food and Agriculture 2002*, FAO, Rome, Italy

FAO (2006) *Reducing Emissions from Deforestation in Developing Countries*, a submission to the UNFCCC, FAO, Rome, Italy

Fausto, B. (1999) *A Concise History of Brazil*, Cambridge University Press, Cambridge, UK

Fearnside, P. (2002) 'Why a 100-year time horizon should be used for global warming mitigation calculations', *Mitigation and Adaptation Strategies for Global Change*, vol 7, pp19–30

Feiden, A., Almeida, D., Vitoi, V. and de Assis, R. (2002) 'Processo de conversão de sistemas de produção orgânicos', *Cadernos de Ciência & Tecnologia (Brasília)* vol 19, no 2, pp179–204 (in Portuguese)

Ferrari, A. and Wall, L. (2004) 'Utilización de árboles fijadores de nitrógeno para la revegetación de suelos degradados', *Revista de la Facultad de Agronomía*, vol 105, no 2, pp63–87 (in Spanish)

Fiori, A., Fiori, C., Disperati, L., Conceição, A., Kozciak, S., Guedes, J. and Ciali, A. (2001) 'O processo erosivo na bacia do Alto Paragui', *Boletim Paranaense de Geociências*, vol 49, pp63–78 (in Portuguese)

Fisher, M. and Thomas, R. (2004) 'Implications of land use change to introduced pastures on carbon stocks in the central lowlands of tropical South America', *Environment, Development and Sustainability*, vol 6, pp111–131

Fleming, J. (1998) *Historical Perspectives on Climate Change*, Oxford University Press, Oxford, UK

Flinn, K., Vellend, M. and Marks, P. (2005) 'Environmental causes and consequences of forest clearance and agricultural abandonment in central New York, USA', *Journal of Biogeography*, vol 32, pp439–452

Follett, R. (2001) 'Soil management concepts and carbon sequestration in cropland soils', *Soil & Tillage Research*, vol 61, pp77–92

Font, M. (1990) *Coffee, Contention and Change*, Basil Blackwell, Cambridge, USA/Oxford, UK

Fourier, J. (1827) 'Memoire sur les temperatures du globe terrestre et des espaces planetaires', *Mémoires de l'Académie Royale des Sciences*, vol 7, pp569–604

Franco, F., Couto, L., Carvalho, A., Jucksch, I. and Fernandes, E. (2002) 'Quantificação de erosãoem sistemas agroflorestais e convencionais na zona da mata de Minas Gerais', *Revista Árvore*, vol 26, no 6, pp751–760 (in Portuguese)

Fresco, L. (2003) *Plant Nutrients: What We Know, Guess and Do Not Know*, Address to the IFA/FAO Agriculture Conference 'Global Food Security and the Role of Sustainable Fertilisation', Rome, Italy, 26–28 March, 2003

Fujisaka, S., Jayson, E. and Dapusala, A. (1994) 'Trees, grasses, and weeds: Species choices in farmed-developed contour hedgerows', *Agroforestry Systems*, vol 25, pp13–22

Galbraith, J. (1987) *A History of Economics*, Hamish Hamilton, London, UK

García-Oliva, F. and Masera, O. (2004) 'Assessment and measurement issues related to soil carbon sequestration in land-use, land-use change, and forestry (LULUCF) projects under the Kyoto Protocol', *Climatic Change*, vol 65, pp347–364

Garrity, D. (1999) 'Contour farming based on natural vegetative strips: Examining the scope for increased food crop production on sloping land in Asia', *Environment, Development and Sustainability*, vol 1, pp323–336

GEF (2003a) *Rio de Janeiro Sustainable Integrated Ecosystem Management in Production Landscapes of the North-Northwestern Fluminense*, GEF project brief, GEF, Washington, DC, USA

GEF (2003b) *Project Executive Summary: Rio de Janeiro Integrated Ecosystem Management in Production Landscapes of the North-Northwestern Fluminense (NNWF)*, GEF Work Council submission, GEF, Washington, DC, USA

GEF (2005a) *Land Management and its Benefits*, Consultation paper prepared for the Scientific and Technical Advisory Panel (STAP) to the Global Environment Facility, GEF, Washington, DC, USA

GEF (2005b) *Rio de Janeiro Sustainable Integrated Ecosystem Management in Production Landscapes of the North-Northwestern Fluminense*, GEF project document, GEF, Washington, DC, USA

Giger, M., Liniger, H. and Critchley, W. (1999) 'Use of direct incentives and profitability of soil and water conservation in Eastern and Southern Africa', in D. Sanders, P. Huszar, S. Sombatpanit, and T. Enters (eds) *Incentives in Soil Conservation: From Theory to Practice*, Science Publishers, Enfield, USA, pp247–274

Giller, K. (2001) *Nitrogen Fixation in Tropical Cropping Systems*, CABI, Wallingford, UK

Giller, K., Witter, E., Corbeels, M. and Tittonell, P. (2009) 'Conservation agriculture and smallholder farming in Africa: The heretics' view', *Field Crops Research*, vol 114, pp23–34

Gisladottir, G. and Stocking, M. (2005) 'Land degradation control and its global environmental benefits', *Land Degradation & Development*, vol 16, pp99–112

Glover, M. (2009) 'Taking biochar to market: Some essential concepts for commercial success', in J. Lehmann and S. Joseph (eds) *Biochar for Environmental Management*, Earthscan, London, UK, pp375–392

Gold Standard (2009) 'Gold Standard version 2.1: Requirements', www.cdmgoldstandard.org/fileadmin/editors/files/6_GS_technical_docs/GSv2.1/GSv2.1_Requirements.pdf, accessed 18 November, 2010

Gomes, M., Borges, S., Franco, I. and Correa, J. (2003) *Tecnologias Apropriadas à Revitalização da Capacidade de Produção de Água de Mananciais da Produção de Água* (Appropriate Technologies for Rehabilitating Water Production of Water-Producing Springs), 33rd Assembleia Nacional da Assemae, Santo André. Anais da 7a Exposição de Experiências Municipais em Saneamento

Goodchild, T. and Jaby El-Haramein, F. (1996) 'A broad spectrum of barley', *ICARDA Caravan*, vol 2, pp4–5

Gore, A. (1992) *Earth in the Balance*, Plume, New York, USA

Grismer, M., O'Geen, A. and Lewis, D. (2006) *Vegetative Filter Strips for Nonpoint Source Pollution Control in Agriculture*, Publication 8195 of the Division of Agriculture and Natural Resources, University of California at Davis, USA

Grubb, M. (2008) *Energy and Climate: Opportunities for the G8*, Climate Strategies, Cambridge, UK

Grubb, M., Vrolijk, C. and Brack, D. (1998) *The Kyoto Protocol*, Earthscan, London, UK

Gruhn, P., Goletti, F. and Yudelman, M. (2000) *Integrated Nutrient Management, Soil Fertility, and Sustainable Agriculture: Current Issues and Future Challenges*, Food, Agriculture and the Environment Discussion Paper 32 of the International Food Policy Institute, Washington, DC, USA

Hamilton, K., Sjardin, M., Peters-Stanley, M. and Marcello, T. (2010) *Building Bridges: State of the Voluntary Carbon Markets 2010*, Bloomberg New Energy Finance, New York, USA

Hardin, G. (1968) 'The tragedy of the commons', *Science*, vol 162, no 13, pp1243–1248

Harrabin, R. (2006) '"£1bn windfall" from carbon trade', BBC News website, 1 May 2006, http://news.bbc.co.uk/2/hi/science/nature/4961320.stm, accessed 12 December 2010

Hart, J. (1968) 'Loss and abandonment of cleared farm land in the Eastern United States', *Annals of the Association of American Geographers*, vol 58, no 3, pp417–440

Helfand, S. and Brunstein, L. (2000) *The Changing Structure of the Brazilian Agricultural Sector and the Limitations of the 1995/96 Agricultural Census*, Paper prepared for the VII NEMESIS Seminar, 7–8 December 2000, Rio de Janeiro, Brazil

Hellin, J. and Haigh, M. (2002) 'Better land husbandry in Honduras: Towards the new paradigm in conserving soil, water and productivity', *Land Degradation & Development*, vol 13, no 3, pp233–250

Herweg, K. (1993) 'Problems of acceptance and adaption of soil conservation in Ethiopia', in E. Baum, P. Wolff and M. Zöbisch (eds) *Acceptance of Soil and Water Conservation: Strategies and Technologies*, DITSL, Witzenhausen, Germany, pp391–411

Hirota, M. (2003) 'Monitoring the Brazilian Atlantic Forest cover', in C. Galindo-Leal and I. Câmara (eds) *The Atlantic Forest of South America – Biodiversity Status, Threats, and Outlook*, Island Press, Washington, DC, USA, pp60–65

Höhne, N., Wartmann, S., Herold, A. and Freibauer, A. (2007) 'The rules for land use, land use change and forestry under the Kyoto Protocol – lessons learned for the future climate negotiations', *Environmental Science & Policy*, vol 10, pp353–369

Holl, K. (1999) 'Factors limiting tropical rainforest regeneration in abandoned pasture: Seed rain, seed germination, microclimate, and soil', *Biotropica*, vol 31, no 2, pp229–242

Hotelling, H. (1931) 'The economics of exhaustible resources', in *Journal of Political Economy*, vol 39, no 2, pp137–175

Houghton, R., Davidson, E. and Woodwell, G. (1998) 'Missing sinks, feedbacks, and understanding the role of terrestrial ecosystems in the global carbon balance', *Global Biogeochemical Cycles*, vol 12, no 1, pp25–34

Hurni, H. (2000) 'Assessing sustainable land management (SLM)', *Agriculture, Ecosystems and Environment*, vol 81, pp83–92

IBGE (1996) *Censo Agropecuário 1995–96*, Instituto Brasileiro de Geografia e Estatística, Brasília, Brazil

IBGE (2002) *Perfil dos Municípios Brasileiros (Meio Ambiente) 2002: Pesquisa de Informações Básicas Municipais*, Instituto Brasileiro de Geografia e Estatística Brasília, Brazil

ICARDA (1995) 'Real plants for real farmers', *ICARDA Caravan*, vol 1, pp6–7

IFAS (2003) *Kyrgyzstan*, Report prepared by the Scientific Information Centre Aral (IFAS) of the Sustainable Development Commission, IFAS, Bishkek, Kyrgyzstan

IFOAM (2005) 'Brazilian government sets sights high for organic agriculture', Press release, 15 December 2005, International Federation of Organic Agriculture Movements (IFOAM), www.ifoam.org/press/press/Government_Brazil_Supports_Organic.html, accessed 24 February 2007

INCRA/FAO (2000) *Novo Retrato da Agricultura Familiar*, INCRA/FAO, Brasília, Brazil (in Portuguese)

IPCC (2000) *Summary for Policymakers, Special Report on Land Use, Land-Use Change, and Forestry*, Cambridge University Press, Cambridge, UK

IPCC (2001a) *Climate Change 2001: The Scientific Basis. Contribution of Working Group I to the Third Assessment Report*, Cambridge University Press, Cambridge, UK

IPCC (2001b) *Special Report on Emissions Scenarios*, Cambridge University Press, Cambridge, UK

IPCC (2001c) *Climate Change 2001: Impacts, Adaptation and Vulnerability. Contribution of Working Group II to the Third Assessment Report of the Intergovernmental Panel on Climate Change,* Cambridge University Press, Cambridge, UK

IPCC (2007a) 'Summary for policymakers', in *Summary of the Report of Working Group I to the Fourth Assessment Report of the Intergovernmental Panel on Climate Change*, Cambridge University Press, Cambridge, UK

IPCC (2007b) 'Food, fibre and forest products', in *Climate Change 2007: Mitigation. Contribution of Working Group III to the Fourth Assessment Report of the Intergovernmental Panel on Climate Change*, Cambridge University Press, Cambridge, UK

IPCC (2007c) 'Ecosystems, their properties, goods, and services', in *Climate Change 2007: Impacts, Adaptation and Vulnerability. Contribution of Working Group II to the Fourth Assessment Report of the Intergovernmental Panel on Climate Change*, Cambridge University Press, Cambridge, UK.

IPGRI (1996) *Access to Plant Genetic Resources and the Equitable Sharing of Benefits: A Contribution to the Debate on Systems for the Exchange of Germplasm*, Issues in Genetic Resources Paper No 4, IPGRI, Rome, Italy

Izac, A. (1997) 'Developing policies for soil carbon management in tropical regions', *Geoderma*, vol 79, pp261–276

Izaurralde, R. and Rice, C. (2006) 'Methods and tools for designing a pilot soil carbon sequestration project', in R. Lal, C. Cerri, M. Bernoux, J. Etchevers, and E. Cerri (eds) *Carbon Sequestration in Soils of Latin America*, Food Products Press, New York, USA, pp457–476

Izaurralde, R., Haugen-Kozyra, K., Jans, D., Mcgill, W., Grant, R. and Hiley, J. (2001) 'Soil C dynamics: Measurement, simulation and site-to-range scale-up', in R. Lal, J. Kimble, R. Follett and B. Stewart (eds) *Assessment Methods for Soil Carbon*, CRC Press, Boca Raton, FL, pp553–575

Jahn, M., Michaelowa, A. and Raubenheimer, S. (2003) *Unilateral CDM – Chances and Pitfalls,* PGZ Eschborn, Germany

Jenkinson, D. (1988) 'Soil organic matter and its dynamics', in A. Wild (ed) *Russell's Soil Conditions and Plant Growth* (11th edn), Longman Scientific and Technical, Harlow, UK, pp564–607

Jones, A., Kumar, P., Saxena, K., Kulkarnia, N., Muniyappa, V. and Waliyar, F. (2004) 'Sterility mosaic disease – the "Green plague" of pigeonpea. Advances in understanding the etiology, transmission and control of a major virus disease', *Plant Disease*, vol 88, no 5, pp436–445

Jones, C., McConnell, C., Coleman, K., Cox, P., Falloon, P., Jenkinson, D. and Powlson, D. (2005) 'Global climate change and soil carbon stocks: Predictions from two contrasting models for the turnover of organic carbon in soil', *Global Change Biology*, vol 11, pp154–166

Jones, J., Koo, J., Naab, J., Bostick, W., Traore, S. and Graham, W. (2007) 'Integrating stochastic models and *in situ* sampling for monitoring soil carbon sequestration', *Agricultural Systems* vol 94, pp52–62

Jones, P. and Thornton, P. (2003) 'The potential impacts of climate change on maize production in Africa and Latin America in 2055', *Global Environmental Change*, vol 13, pp51–59

Jung, T., Srinavasan, A., Tamura, K., Sudo, T., Watanabe, R., Shimada, K. and Kimura, H. (2005) *Asian Perspectives on Climate Regime Beyond 2012: Concerns, Interests and Priorities*, Institute for Global Environmental Strategies, Hayama, Japan

Karousakis, K. (2007) *Incentives to Reduce GHG Emissions from Deforestation: Lessons Learned from Costa Rica and Mexico*, OECD/International Energy Agency, Paris, France

Kelliher, F. and Clark, H. (2010) 'Ruminants', in D. Reay, P. Smith and A. van Amstel, *Methane and Climate Change*, Earthscan, London, UK, pp136–150

Kerckhoffs, L. and Reid, J. (2007) *Carbon Sequestration in the Standing Biomass of Orchard Crops in New Zealand*, Report prepared for Horticulture New Zealand Ltd., New Zealand Institute for Crop & Food Research Ltd., Hastings, New Zealand

Kerr, J. and Sanghi, N. (1993) 'Indigenous soil and water conservation in India's semi-arid tropics', in E. Baum, P. Wolff and M. Zöbisch (eds) *Acceptance of Soil and Water Conservation: Strategies and Technologies*, DITSL, Witzenhausen, Germany, pp255–290

Kerr, J., Pangare, G., Pangare, V. and George, P. (1999) 'Effects of watershed project subsidies on soil and water conservation investments in India's semi-arid tropics', in D. Sanders, P. Huszar, S. Sombatpanit and T. Enters (eds) *Incentives in Soil Conservation: From Theory to Practice*, Science Publishers, Enfield, USA, pp295–308

Kimble, J. (2006) 'Advances in models to measure soil carbon: Can soil carbon really be measured?', in R. Lal, C. Cerri, M. Bernoux, J. Etchevers and E. Cerri (eds) *Carbon Sequestration in Soils of Latin America*, Food Products Press, Binghampton, NY, USA, pp477–490

Kimble, J., Grossman, R. and Samson-Liebig, S. (2001) 'Methods for assessing soil carbon pools', in R. Lal, J. Kimble, R. Follett and B. Stewart (eds) *Assessment Methods for Soil Carbon*, CRC Press, Boca Raton, USA, pp3–15

Kirschbaum, M. (2000) 'Will changes in soil organic carbon act as a positive or negative feedback on global warming?', *Biogeochemistry*, vol 48, pp21–51

Kolk, A. (1998) 'From conflict to cooperation: International policies to protect the Brazilian Amazon', *World Development*, vol 26, no 8, 1481–1493

Kossoy, A. and Ambrosi, P. (2010) *State and Trends of the Carbon Market 2010*, The World Bank, Washington, DC, USA

Kotschi, J. and Müller-Sämann, K. (2004) *The Role of Organic Agriculture in Mitigating Climate Change*, Scoping study prepared for the International Federation of Organic Agriculture Movements, Bonn, Germany

Kuhn, N., Hoffmann, T., Schwanghart, W. and Dotterweich, M. (2009) 'Agricultural soil erosion and global carbon cycle: Controversy over?', Letter in *Earth Surface Processes and Landforms*, vol 34, pp1033–1038

La Rovere, E. and Pereira, A. (2005) 'Brazil and climate change: A country profile', SciDev.net, www.scidev.net/dossiers/index.cfm?fuseaction=policybrief&policy=88&dossier=4, accessed 12 February 2007

Laing, D. and Ashby, J. (1993) 'Adoption of improved land management practices by resource-poor farmers', in E. Baum, P. Wolff and M. Zöbisch (eds) *Acceptance of Soil and Water Conservation: Strategies and Technologies*, DITSL, Witzenhausen, Germany, pp59–76

Lal, R. (1997) 'Residue management, conservation tillage and soil restoration for mitigating greenhouse effect by CO_2 enrichment', *Soil and Tillage Research*, vol 43, pp81–107

Lal, R. (2001)'Soil degradation by erosion', *Land Degradation & Development*, vol 12, no 6, pp519–539

Lal, R. (2002)'Soil carbon dynamics in cropland and rangeland', *Environmental Pollution*, vol 116, pp353–362

Lal, R. (2003) 'Soil erosion and the global carbon budget', *Environment International*, vol 29, no 4, pp437–450

Lal, R. (2004a) 'Soil carbon sequestration impacts on global climate change and food security', *Science*, vol 304, pp1623–1627

Lal, R. (2004b) 'Offsetting China's CO_2 emissions by soil carbon sequestration', *Climatic Change*, vol 65, pp263–275

Lal, R. (2004c) 'Soil carbon sequestration in India', *Climatic Change*, vol 65, pp277–296

Lal, R. (2006) 'Carbon management in agricultural soils', *Mitigation and Adaptation Strategies for Global Change*, vol 12, no 2, pp303–322

Lal, R. (2009) 'Challenges and opportunities in soil organic matter research', *European Journal of Soil Science*, vol 60, pp158–169

Lal, R. and Bruce, J. (1999) 'The potential of world cropland soils to sequester C and mitigate the greenhouse effect', *Environmental Science & Policy*, vol 2, pp177–185

Lal, R. and Kimble, J. (1997) 'Conservation tillage for carbon sequestration', *Nutrient Cycling in Agroecosystems*, vol 49, pp243–253

Lal, R. and Pimentel, D. (2008) 'Soil erosion: A carbon sink or source?', *Science*, vol 319, pp1040–1042

Lecoq, F. and Capoor, K. (2005) *State and Trends of the Carbon Market 2005*, The World Bank, Washington, DC, USA

Lehmann, J. and Joseph, S. (2009) 'Biochar for environmental management: An introduction', in J. Lehmann and S. Joseph (eds) *Biochar for Environmental Management*, Earthscan, London, UK, pp1–12

Lehmann, J., Amonetee, J. and Roberts, K. (2010) 'Role of biochar in mitigation of climate change', in D. Hillel and C. Rosenzweig (eds) *Handbook of Climate Change and Agroecosystems: Impacts, Adaptation, and Mitigation*, Imperial College Press, London, UK, pp343–363

Levinson, D., Hilburn, K., Lawrimore, J. and Kruk, M. (2010) 'Global precipitation', in D. Arndt, M. Baringer and M. Johnson (eds) *State of the Climate in 2009*, Special Supplement to the Bulletin of the American Meteorological Society, vol 91, pp31–32

Lewis, L. (1992) 'Terracing and accelerated soil loss on Rwandian steeplands: A preliminary investigation of the implications of human activities affecting soil movement', *Land Degradation & Rehabilitation*, vol 3, no 4, pp241–246

Lewis, M. (1992) *Green Delusions: An Environmentalist Critique of Radical Environmentalism*, Duke University Press, Durham, USA

Loisel, C. (2008) *Climate Change Mitigation in the Forest Sector: What Happened in Poznan*, Policy Brief No. 10/2008 of the Institute for Sustainable Development and International Relations, IDDRI, Paris, France

Lopes, M., Lima, A., Carvalho, F., Reis, R., Santos, Í. and Saraiva, F. (2004) 'Controle gerencial e estudo da rentibilidade de sistemas de produção de leite na região de Lavras (MG)', *Ciência e Agrotecnologia*, vol 28, no 4, pp883–892 (in Portuguese)

Lowdermilk, W. (1948) *Conquest of the Land through Seven Thousand Years*, United States Department of Agriculture Soil Conservation Service, Washington, USA

Lu, Y. and Stocking, M. (2000a) 'Integrating biophysical and socio-economic aspects of soil conservation on the Loess Plateau, China. Part II: Productivity impact and economic costs of erosion', *Land Degradation & Development*, vol 11, no 2, pp141–152

Lu, Y. and Stocking, M. (2000b) 'Integrating biophysical and socio-economic aspects of soil conservation on the Loess Plateau, China. Part III: The benefits of conservation', *Land Degradation & Development*, vol 11, no 2, pp153–166

Ludwig, B. and Khanna, P. (2001) 'Use of near infrared spectroscopy to determine inorganic and organic carbon fractions in soil and litter', in R. Lal, J. Kimble, R. Follett and B. Stewart (eds) *Assessment Methods for Soil Carbon*, CRC Press, Boca Raton, USA, pp361–370

Maletta, H. (2000) *Brazilian Agriculture Towards 2020*, Facultad de Ciencias Sociales, Universidad del Salvador, Buenos Aires, Argentina

Markandya, A. (1991) 'Global warming: The economics of tradeable permits', in D. Pearce (ed) *Blueprint 2: Greening the World Economy*, Earthscan Publications/London Environmental Economics Centre, London, UK, pp53–74

Martens, P., Kobats, R., Nijhof, S., de Vries, P., Livermore, M., Bradley, D., Cox, J. and McMichael, A. (1999) 'Climate change and future populations at risk of malaria', *Global Environmental Change*, vol 9, S89–S107

Matos, R. and Giovanini, R. (2004) 'Geohistória econômica da Zona da Mata Mineira', in João Antonio de Paula et al (eds) *Anais do XI Seminário sobre a Economia Mineira*, Centro de Desenvolvimento e Planejamento Regional (Cedeplar), Universidade Federal de Minas Gerais, Belo Horizonte, Brazil (in Portuguese)

Matthey, H., Fabiosa, J. and Fuller, F. (2004) *Brazil: The Future of Modern Agriculture?* MATRIC Briefing Paper 04-MBP 6, Midwest Agribusiness Trade Research and Information Center, Iowa State University Ames, Iowa, USA

May, P. and Geluda, L. (2004) *Pagamentos para Serviços Ecossistêmicos*, Projeto Gerenciamento Integrado de Agroecossistemas em Microbacias Hidrográficas no Norte/Noroeste Fluminense – Rio Rural/GEF-BIRD/SEAAPI-RJ Rio de Janeiro, Brazil (in Portuguese)

May, P. H., Boyd, E., Veiga, F. and Chang, M. (2004) *Local Sustainable Development Effects of Forest Carbon Projects in Brazil and Bolivia: A View from the Field*, IIED, London, UK

McCarl, B. and Murray, B. (2001) *Harvesting the Greenhouse: Comparing Biological Sequestration with Emissions Offsets*, Texas A&M University, College Station, USA

McCarl, B., Peacocke, C., Chrisman, R., Kung, C. and Sands, R. (2009) 'Economics of biochar production, utilization and greenhouse gas offsets', in J. Lehmann and S. Joseph (eds) *Biochar for Environmental Management*, Earthscan, London, UK, pp341–358

McDonagh, J., Birch Thomsen, T. and Magid, J. (2001) 'Soil organic matter decline and compositional change associated with cereal cropping in southern Tanzania', *Land Degradation & Development*, vol 12, no 1, pp13–26

Meinshausen, M. and Hare, B. (2002) *Temporary Sinks Do Not Cause Permanent Climatic Benefits: Achieving Short-Term Emission Reduction Targets at the Future's Expense*, Greenpeace Background Paper, Greenpeace International, Amsterdam, The Netherlands

Mello, F., Cerri, C., Bernoux, M., Volkoff, B. and Cerri, C. (2006) 'Potential for carbon sequestration in soils of the Atlantic Forest Region of Brazil', in R. Lal, C. Cerri, M. Bernoux, J. Etcheversand E. Cerri (eds) *Carbon Sequestration in Soils of Latin America*, Food Products Press, Binghampton, NY, USA, pp349–368

Mendelsohn, R. and Williams, L. (2004) 'Comparing forecasts of the global impacts of climate change', *Mitigation and Adaptation Strategies for Global Change*, vol 9, pp315–333

Meszaros, G. (2000) 'No ordinary revolution: Brazil's landless workers' movement', *Race and Class*, vol 42, no 2, pp1–18

Meyer, C. (2007) 'The FO: Smug, timid, conformist', *The Sunday Times*, Culture Section, 4 March 2007, pp46–47

Mitford, J. (1960) *Hons and Rebels*, Victor Gollancz, London, UK

Mooney, H., Roy, J. and Saugier, B. (eds) (2001) *Terrestrial Global Productivity: Past, Present and Future*, Academic Press, San Diego, USA

Morgan, J., Milchunas, D., LeCain, D., West, M. and Mosier, A. (2007) 'Carbon dioxide enrichment alters plant community structure and accelerates shrub growth in the shortgrass steppe', *Proceedings of the National Academy of Science*, vol 104, no 37 pp14724–14729

Morgan, P., Bollero, G., Nelson, R., Dohleman, F. and Long, S. (2005) 'Smaller than predicted increase in aboveground net primary production and yield of field-grown soybean under fully open-air [CO_2] elevation', *Global Change Biology*, vol 11, pp1856–1865

Morison, J., Hine, R. and Pretty, J. (2005) 'Survey and analysis of labour on organic farming in the UK and Republic of Ireland', *International Journal of Agricultural Sustainability*, vol 3, no 1, pp24–43

Mrabet, R., Ibno-Namr, K., Bessam, F. and Saber, N. (2001) 'Soil chemical quality changes and implications for fertilizer management after 11 years of no-tillage wheat production systems in semiarid Morocco', *Land Degradation & Development*, vol 12, no 6, pp505–518

Muchagata, M. and Brown, K. (2003) 'Cows, colonists and trees: Rethinking cattle and environmental degradation in Brazilian Amazonia', *Agricultural Systems*, vol 76, pp797–816

Müller, B. (2007) *Bonn 2007: Russian Proposals, Policy CDM, and 'CER Put Options' (CERPOs)*, OIES Energy and Environment Comment, July 2007

Najam, A., Huq, S. and Sokona, Y. (2003) 'Climate negotiations beyond Kyoto: Developing countries concerns and interests', *Climate Policy*, vol 3, pp221–231

Nasyrov, M. (2000) 'Global warming and the rangelands', *ICARDA Caravan*, vol 13, pp31–32

Nelson, K. and de Jong, B. (2003) 'Making global initiatives local realities: Carbon mitigation projects in Chiapas, Mexico', in *Global Environmental Change*, vol 13, pp19–30

Neves, M., Medeiros, C., Almeida, D., De-Polli, H., Rodrigues, H., Guerra, J., Nunes, M., Cardoso, M., Azevedo, M., Vieira, R. and Saminêz, T. (2000) *Agricultura Orgânica: Instrumento para a Sustentabilidade dos Sistemas de Produção e Valoração de Produtos Agropecuários*, Publication 122, EMBRAPA Agrobiología, Seropédica, RJ, Brazil (in Portuguese)

Neves, M., De-Polli, H., Peixoto, R. and Almeida, D. (2002) 'Por que não utilizar uréia como fontre de N na agricultura orgânica', *Cadernos de Ciência & Tecnologia (Brasília)*, vol 19, no 2, pp313–331 (in Portuguese)

Nielsen, T. and Zöbisch, M. (2001) 'Multi-factorial causes of land-use change: Land-use dynamics in the agropastoral village of Im Mial, Northwestern Syria', *Land Degradation & Development*, vol 12, no 2, 143–162

Niles, J., Brown, S., Pretty, J., Ball, S. and Fay, J. (2003) 'Potential carbon mitigation and income in developing countries from changes in use and management of agricultural and forest lands', in I. Swingland (ed) *Capturing Carbon and Conserving Biodiversity: The Market Approach*, Earthscan, London, UK, pp70–89

Noble, I. and Scholes, R. (2001) 'Sinks and the Kyoto Protocol', *Climate Policy*, vol 1, pp5–25

Norgaard, R. and Bode, C. (1998) 'Next, the value of God, and other reactions', *Ecological Economics,* vol 25, pp37–39

Norse, D. (2003) *Fertilizers and World Food Demand, Implications for Environmental Stresses*, Paper presented to the IFA/FAO Agriculture Conference 'Global Food Security and the Role of Sustainable Fertilisation', Rome, Italy, 26–28 March 2003

Open Europe (2007) *Europe's Dirty Secret: Why the EU Emissions Trading Scheme Isn't Working*, Open Europe, London, UK

Pádua, J. (1999) 'Aniquilando as naturais produções: Crítica iluminista, crise colonial e as origens do ambientalismo político no Brasil (1786–1810)', *Dados*, vol 42, no 3 (in Portuguese)

Pádua, J. (2004) *Nature Conservation and Nature Building in the Thought of a Brazilian Founding Father: José Bonifácio (1763–1838),* Working Paper no CBS-53-04, Centre for Brazilian Studies, Oxford, UK

Pagiola, S. (1999a) *The Global Environmental Benefits of Land Degradation Control on Agricultural Land*, World Bank Environment Paper 16, The World Bank, Washington, DC, USA

Pagiola, S. (1999b) 'Economic analysis of incentives for soil conservation', in D. Sanders, P. Huszar, S. Sombatpanit and T. Enters (eds) *Incentives in Soil Conservation: From Theory to Practice*, Science Publishers, Enfield, USA, pp41–56

Panda Standard (2009) *Panda Standard v1.0*, Panda Standard, Beijing, China

Parry, M., Rosenzweig, C., Iglesias, A., Fischer, G. and Livermore, M. (1999) 'Climate change and world food security: A new assessment', *Global Environmental Change*, vol 9, S51–S67

Parry, M., Rosenzweig, C., Iglesias, A., Livermore, M. and Fischer, G. (2004) 'Effects of climate change on global food production under SRES emissions and socio-economic scenarios', *Global Environmental Change*, vol 14, pp53–67

Pearce, D., Markandya, A. and Barbier, E. (1989) *Blueprint for a Green Economy*, Earthscan, London, UK

Pimentel, D. (2006) 'Soil erosion: A food and environmental threat', *Environment, Development and Sustainability*, vol 8, pp119–137

Pirard, R. and Treyer, S. (2010) *Agriculture and Deforestation: What Role Should REDD+ and Public Support Policies Play?* IDDRI, Paris, France

Pohl, O. (2001) 'A harvest of bounty and woe', in *Christian Science Monitor*, 22 August 2001, www.csmonitor.com/2001/0822/p6s1-woeu.html, accessed 2 August, 2007

Post, W., Izaurralde, R., Mann, L. and Bliss, N. (2001) 'Monitoring and verifying changes of organic carbon in soil', *Climatic Change*, vol 51, pp73–99

Poussart, J-N., Ardö, J. and Olsson, L. (2004a) 'Verification of soil carbon sequestration: Sample requirements', *Environmental Management*, vol 33 (Supplement 1), S416–S425

Poussart, J-N., Ardö, J. and Olsson L. (2004b) 'Effects of data uncertainties on estimated soil organic carbon in the Sudan', *Environmental Management*, vol 33 (Supplement 1), S405–S415

Pretty, J. and Ball, A. (2001) 'Agricultural influences on carbon emissions and sequestration: A review of evidence and the emerging trading options', University of Essex, Centre for Environment and Society, Occasional Paper 2001-03, University of Essex, Colchester, UK

Pretty, J. and Shah, P. (1999) 'Soil and water conservation: A brief history of coercion control', in F. Hinchcliffe, J. Pretty, and J. Thompson (eds) *Fertile Ground: The Impacts of Participatory Watershed Management*, Intermediate Technology Publications, Rugby, UK, pp1–12

Pretty, J. and Shah, P. (1997) 'Making soil and water conservation sustainable: From coercion and control to partnerships and participation', *Land Degradation & Development*, vol 8, no 1, pp39–58

Pretty, J., Noble, A., Bossio, D., Dixon, J., Hine, R., Penning de Vries, F. and Morrison, J. (2006) 'Resource-conserving agriculture increases yields in developing countries', *Environmental Science and Technology*, vol 40, no 4, pp1114–1119

Primavesi, A. and Primavesi, O. (2002) 'Optimising climate-soil-pasture-cattle interactions in Brazil', *Leisa Magazine*, April 2002, pp12–13

Prudencio, C. (1993) 'Ring management of soils and crops in the West African semi-arid tropics: The case of the Mossi farming system in Burkina Faso', *Agriculture, Ecosystem and Environment*, vol 47, pp237–264

Raphael, D. (1997) 'Smith', in D. Raphael, D. Winch and R. Skidelsky, *Three Great Economists*, Oxford University Press, Oxford, UK, pp1–105

Reardon, T., Berdegué, J. and Escobar, G. (2001) 'Rural nonfarm employment and incomes in Latin America: Overview and policy implications', *World Development*, vol 29, no 3, pp395–409

Rees, W. (1998) 'How should a parasite value its host?', *Ecological Economics*, vol 25, pp49–52

Reeves, J., McCarty, G., Follett, R. and Kimble, J. (2006) 'The potential of spectroscopic methods for the rapid analysis of soil samples', in R. Lal, C. Cerri, M. Bernoux, J. Etchevers and E. Cerri (eds) *Carbon Sequestration in Soils of Latin America*, Food Products Press, Binghampton, NY, USA, pp423–442

Ringius, L. (1999) *Soil Carbon Sequestration and the CDM: Opportunities and Challenges for Africa*, prepared for UNEP Collaborating Centre on Energy and Environment (UCCEE) and Center for International Climate and Environmental Research – Oslo (CICERO), UCCEE/CICERO, Roskilde, Denmark, and Oslo, Norway

Ringius, L. (2002) 'Soil carbon sequestration and the CDM: Opportunities and Challenges for Africa', *Climatic Change*, vol 54, pp471–495

Robbins, M. (2004) *Carbon Trading, Agriculture and Poverty*, Special Publication No. 2 of the World Association of Soil and Water Conservation, WASWC, Bangkok, Thailand

Roberts, T. (2006) 'Some reflections on Brazilian agriculture', *Tropical Agriculture Association Newsletter*, pp24–26

Robertson, G. and Grace, P. (2004) 'Greenhouse gas fluxes in tropical and temperate agriculture: The need for a full-cost accounting of global warming potentials', *Environment, Development and Sustainability*, vol 6, pp51–63

Rosa, H., Kandel, S., Dimas, S. and Dimas, L. (2004) *Payment for Environmental Services and Rural Communities: Lessons from the Americas*, Working Paper 96, Political Economy Research Institute, University of Massachusetts at Amherst, USA

Rosenberg, N., Izaurralde, R. and Malone, E. (eds) (1999) *Carbon Sequestration in Soils: Science, Monitoring and Beyond,* proceedings of the St Michaels Workshop, December 1998, Battelle Press, Columbus, USA

Rudel, T. (2009) 'Reinforcing REDD+ with reduced emissions agricultural policy', in A. Angelsen (ed) *Realising REDD+: National Strategy and Policy Options*, CIFOR, Bogor, Indonesia, pp191–200

Russell, B. (1946) *A History of Western Philosophy*, George Allen & Unwin, London, UK

Salzman, J., Thompson, B. and Daily, G. (2001)'Protecting ecosystem services: Science, economics, and policy', *Stanford Environmental Law Journal*, vol 20, no 2, pp309–332

Sample, I. (2005) 'Warming hits "tipping point"', *Guardian*, 11 August 2005

Sanabria, D., Silva-Acuña, R. and Marcano, M. (2006) 'Evaluación de tres sistemas de labranza en la recuperación de una pastura degradada', *Zootecnia Tropical*, vol 24, no 4, pp417–433 (in Spanish)

Sanders, D. and Cahill, D. (1999) 'Where incentives fit in soil conservation programs', in D. Sanders, P. Huszar, S. Sombatpanit and T. Enters (eds) *Incentives in Soil Conservation: From Theory to Practice*, Science Publishers, Enfield, USA, pp11–24

Sandford, S. (1983) *Pastoral Development in the Third World*, Wiley and Son, London, UK.

Scherr, S. (2001) 'The future food security and economic consequences of soil degradation in the developing world', in E. Bridges, I. Hannam, L. Oldeman, F. Penning de Vries,

S. Scherr and S. Sombatpanit (eds) *Response to Land Degradation*, Science Publishers Inc., Enfield, USA, pp155–170

Scherr, S., White, A. and Khare, A (2004) *For Services Rendered: The Current Status and Future Potential of Markets for the Ecosystem Services Provided by Tropical Forests*, ITTO Technical Series No 21, International Tropical Timber Organisation, Yokohama, Japan

Schlamadinger, B., Johns, T., Ciccarese, L., Braun, M., Sato, A., Senyaz, A., Stephens, P., Takahashi, M. and Zhang, X. (2007a) 'Options for including land use in a climate agreement post-2012: Improving the Kyoto Protocol approach', *Environmental Science & Policy*, vol 10, pp295–305

Schlamadinger, B., Bird, N., Johns, T., Brown, S., Canadell, J., Ciccarese, L., Dutschke, M., Fiedler, J., Fischlin, A., Fearnside, P., Forner, C., Freibauer, A., Frumhoff, P., Höhne, N., Kirschbaum, M., Labat, A., Marland, G., Michaelowa, A., Montanarella, L., Moutinho, P., Murdiyarso, D., Pena, N., Pingoud, K., Rakonczay, Z., Rametsteiner, E., Rock, J., Sanz, M., Schneider, U., Shvidenko, A., Skutsch, M., Smith, P., Somogyi, Z., Trines, E., Ward, M. and Yamagata, Y. (2007b) 'A synopsis of land use, land-use change and forestry (LULUCF) under the Kyoto Protocol and Marrakech Accords', *Environmental Science & Policy*, vol 10, pp271–282

Schlosser, J., Debiasi, H., Parcianello, G. and Rambo, L. (2002) 'Caracterização dos acidentes com tratores agrícolas', *Ciencia Rural*, vol 6, pp977–981 (in Portuguese)

Scholes, B. (1999) 'Will the terrestrial carbon sink saturate soon?', *Global Change Newsletter*, no 37, pp2–3

Scripps Institution (2005) 'Roger Randall Dougan Revelle biography', Scripps Institution of Oceanography Archives, http://scilib.ucsd.edu/sio/archives/siohstry/revelle-biog.html, accessed 12 November 2005

Seeberg-Elverfeldt, C. and Tapio-Biström, M. (2010) *Global Survey of Agricultural Mitigation Projects*, Mitigation of Climate Change in Agriculture Project (MICCA), FAO, Rome, Italy

Seymour, F. and Angelsen, A. (2009) 'Summary and conclusions: REDD wine in old wineskins?', in A. Angelsen (ed) *Realising REDD+: National Strategy and Policy Options*, CIFOR, Bogor, Indonesia, pp293–304

Shackley, S. and Sohi, S. (eds) (2010) *An Assessment of the Benefits and Issues Associated with the Application of Biochar to Soil*, Report commissioned by the United Kingdom Department for the Environment, Food and Rural Affairs and the Department of Energy and Climate Change, UK Biochar Research Centre, Edinburgh, UK

Sills, E., Sunderlin, W. and Wertz-Kanounnikoff, S. (2009) 'The evolving landscape of REDD+ projects', in A. Angelsen (ed) *Realising REDD+: National Strategy and Policy Options*, CIFOR, Bogor, Indonesia, pp265–280

Silvano, R., Udvardy, S., Ceroni, M. and Farley, J. (2005) 'An ecological integrity assessment of a Brazilian Atlantic Forest watershed based on surveys of stream health and local farmers' perceptions: Implications for management', *Ecological Economics*, vol 53, pp369–385

Silver, W., Ostertag, R. and Lugo, A. (2000) 'The potential for carbon sequestration through reforestation of abandoned tropical agricultural and pasture lands', *Restoration Ecology*, vol 8, no 4, pp394–407

Sisti, C., dos Santos, H., Kohhan, R., Alves, B., Urquiaga, S. and Boddey, R. (2004) 'Change in carbon and nitrogen stocks in soil under 13 years of conventional or zero tillage in southern Brazil', *Soil & Tillage Research*, vol 76, pp39–58

Six, J. and Jastrow, J. (2002) 'Organic matter turnover', in R. Lal (ed) *Encyclopedia of Soil Science*, Marcel Dekker, Inc., New York, USA, pp936–942

Six, J., Conant, R., Paul, E. and Paustian, K. (2000) 'Stabilization mechanisms of soil organic matter: Implications for C-saturation of soils', *Plant and Soil*, vol 241, pp155–176

Six, J., Bossuyt, H., Degryze, S. and Denef, K. (2004) 'A history of research on the link between (micro)aggregates, soil biota, and soil organic matter dynamics', *Soil and Tillage Research*, vol 79, pp7–31

Smith, A. (1776) *An Inquiry into the Nature and Causes of the Wealth of Nations*, W. Strahan and T. Cadell, London

Smith, P. (2004) 'Monitoring and verification of soil carbon changes under Article 3.4 of the Kyoto Protocol', *Soil Use and Management*, vol 20, pp264–270

Smith, P., Martino, D., Cai, Z., Gwary, D., Janzen, H., Kumar, P., McCarl, B., Ogle, S., O'Mara, F., Rice, C., Scholes, B., Sirotenko, O., Howden, M., McAllister, T., Pan, G., Romanenkov, V., Schneider, U. and Towprayoon, S. (2007) 'Policy and technological constraints to implementation of greenhouse gas mitigation options in agriculture', *Agriculture, Ecosystems and Environment*, vol 118, pp6–28

Snapp, S., Mafongoya, P. and Waddington, S. (1998) 'Organic matter technologies for integrated nutrient management in smallholder cropping systems of southern Africa', *Agriculture, Ecosystems and Environment*, vol 71, pp185–200

Sombatpanit, S., Sangsingkeo, S., Palasuwan, N. and Saengvichien, S. (1993) 'Soil conservation and farmer's acceptance in Thailand', in E. Baum, P. Wolff and M. Zöbisch (eds) *Acceptance of Soil and Water Conservation: Strategies and Technologies*, DITSL, Witzenhausen, Germany, pp307–340

Sombroek, W., Nachtergaele, F. and Hebel, A. (1993) 'Amounts, dynamics and sequestering of carbon in tropical and subtropical soils', *Ambio*, vol 22, pp417–426

Sombroek, W., Kern, D., Rodrigues, T., Cravo, M., Jarbas, T., Woods, W. and Glaser, B. (2002) *Terra Preta and Terra Mulata: Pre-Columbian Amazon Kitchen Middens and Agricultural Fields, Their Sustainability and Their Replication*, Paper presented at the 17th World Congress of Soil Science, Bangkok, Thailand

Sotomayor-Ramírez, D., Espinoza, Y. and Rámirez-Santana, R. (2006) 'Short-term tillage practices on soil organic matter pools in a tropical Ultisol', *Australian Journal of Soil Research*, vol 44, no 7, pp687–693

Stallard, R. (1998) 'Terrestrial sedimentation and the carbon cycle: Coupling weathering and erosion to carbon burial', *Global Biogeochemical Cycles*, vol 12, pp231–257

Stark, W. (1944) *The History of Economics in its Relation to Social Development*, Kegan Paul, Trench, Trubner & Co. Ltd., London, UK

Stark, M., Garrity, D., Mercado, A. and Jutzi, S. (2000) *Building Research on Farmers' Innovations: Low-Cost Natural Vegetative Strips and Soil Fertility Management*, Paper presented at the Environmental Education Network of the Philippines (EENP) conference at Misamis Oriental State College of Agriculture (MOSCAT), Claveria, Philippines, 31 May to 1 June 2000

Stein, S. (1957) *Vassouras*, Harvard University Press, Cambridge, MA, USA

Stocking, M. (1993) 'Soil and water conservation for resource-poor farmers: Designing acceptable technologies for rainfed conditions in eastern India', in E. Baum, P. Wolff and M. Zöbisch (eds) *Acceptance of Soil and Water Conservation: Strategies and Technologies*, DITSL, Witzenhausen, Germany, pp291–205

Stocking, M. (1996) 'Soil erosion', in W. Adams, A. Goudie and A. Orme (eds) *The Physical Geography of Africa*, Oxford University Press, Oxford, UK, pp326–341

Stroosnijder, L. (2005) 'Measurement of erosion: Is it possible?', *Catena*, vol 64, pp162–173

Sugiyama, T. and Michaelowa, A. (2001) 'Reconciling the CDM with inborn paradox of additionality concept', *Climate Policy*, vol 1, pp75–83

Sumberg, J. and Okali, C. (1997) *Farmers' Experiments: Creating Local Knowledge*, Lynne Reiner Publishers, Inc., Boulder, USA

Sunderlin, W. and Atmadja, S. (2009) 'Is REDD+ an idea whose time has come and gone?', in A. Angelsen (ed) *Realising REDD+: National Strategy and Policy Options*, CIFOR, Bogor, Indonesia, pp45–56

Swallow, B., van Noordwijk, M., Dewi, S., Murdiyarso, M., White, D., Gockowski, J., Hyman, G., Budidarsono, S., Robiglio, V., Meadu, V., Ekadinata, A., Agus, F., Hairiah, K., Mbile, P., Sonwa, D. and Weise, S. (2007) *Opportunities for Avoided Deforestation with Sustainable Benefits: An Interim Report by the ASB Partnership for the Tropical Forest Margins*, ASB Partnership for the Tropical Forest Margins, Nairobi, Kenya

Tabarelli, M., Pinto, L., de Silva, J. and Costa, C. (2003) 'Endangered species and conservation planning', in C. Galindo-Leal and I. Câmara (eds) *The Atlantic Forest of South America – Biodiversity Status, Threats, and Outlook*, Island Press, Washington, DC, USA, pp86–94

Tabarelli, M., Pinto, L., Silva, J., Hirota, M. and Bedê, L. (2005) 'Challenges and opportunities for biodiversity conservation in 'the Brazilian Atlantic Forest', *Conservation Biology*, vol 19, no 3, pp695–700

Tagliari, P. (2003) 'Tomate em plantio direto: Menos agrotóxico, mais renda e mais saúde', *Agropecuária Catarinense*, vol 16, no 3, pp24–29 (in Portuguese)

Taiyab, N. (2006) *Exploring the Market for Voluntary Carbon Offsets*, International Institute for Environment and Development, London, UK

Tarré, R., Macedo, R., Cantarutti, R., Rzende, C., Pereira, J., Ferreira, E., Alves, B., Urquiaga, S. and Boddey, R. (2001) 'The effect of the presence of a forage legume on nitrogen and carbon levels in soils under *Brachiaria* pastures in the Atlantic Forest region of the South of Bahia, Brazil', *Plant and Soil*, vol 234, pp15–26

Thies, J. and Rillig, M. (2009) 'Characteristics of biochar: Biological properties', in J. Lehmann and S. Joseph (eds) *Biochar for Environmental Management*, Earthscan, London, UK, pp85–108

Tiffen, M. (1996) 'Land & capital: Blind spots in the study of the "resource-poor" farmer', in M. Leach and R. Mearns (eds) *The Lie of the Land: Challenging Received Wisdom on the African Environment*, The International African Institute/James Currey Ltd., Oxford/London, UK, pp168–185

Tipper, R. (2002) 'Helping indigenous farmers participate in the international market for carbon services: The case of Scolel Té', in S. Pagiola, J. Bishop and N. Landell-

Mills (eds) *Selling Forest Environmental Services: Market-Based Mechanisms for Conservation*, Earthscan, London, UK, pp223–234

Tompkins, E. (2003) *Using Stakeholders' Preferences in Multi-Attribute Decision Making: Elicitation and Aggregation Issues*, CSERGE Working Paper ECM 03-13, CSERGE/The Tyndall Centre for Climate Change Research, Norwich, UK

Topik, S. (1987) *The Political Economy of the Brazilian State, 1889–1930*, University of Texas Press, Austin, USA

Tribunal de Contas do Estado do Rio de Janeiro (2002) *Estudo Socioeconômico 1997–2001 Seropédica*, Secretaria-Geral de Planejamento, Rio de Janeiro, Brazil (in Portuguese)

Tschakert, P. (2004) 'The costs of soil carbon sequestration: An economic analysis for small-scale farming systems in Senegal', *Agricultural* Systems, vol 81, pp227–253

Tschakert, P. (2006) 'Views from the vulnerable: Understanding climatic and other stressors in the Sahel', *Global Environmental Change*, vol 17, pp381–396

Tschakert, P. (2007) 'Environmental services and poverty reduction: Options for smallholders in the Sahel', *Agricultural Systems*, vol 94, pp75–86

Tschakert, P. and Tappan, G. (2004) 'The social context of carbon sequestration: Considerations from a multi-scale environmental history of the Old Peanut Basin of Senegal', *Journal of Arid Environments*, vol 59, pp535–564

Tschakert, P., Coomes, O. and Potvin, C. (2007) 'Indigenous livelihoods, slash-and-burn agriculture, and carbon stocks in Eastern Panama', *Ecological Economics*, vol 60, pp807–820

Tu, M. (2002) 'Element stewardship abstract for *Cenchrus ciliaris*', Wildland Invasive Species Team, University of California, Davis

Turner, R., Adger, W. and Brouwer R. (1998) 'Ecosystem services value, research needs, and policy relevance: A commentary', *Ecological Economics*, vol 25, pp61–65

UNCCD (2008) *Use of Biochar (Charcoal) to Replenish Carbon Pools, Restore Soil Fertility and Sequester CO_2*, Submission by the United Nations Convention to Combat Desertification to the Ad Hoc Working Group on Long-term Cooperative Action under the Convention (AWG-LCA 4), Poznan, 1–10 December 2008

UNFCCC (2007) *Decision 1/CP.13 (Bali Action Plan)*, downloaded on August 15 2008 from http://unfccc.int/resource/docs/2007/cop13/eng/06a01.pdf#page=3.

Urquiaga, S. and Malavolta, E. (2002) 'Uréia: Um adubo orgânico de potential para a agricutura orgânica', *Cadernos de Ciência & Tecnologia (Brasília)*, vol 19, no 2, pp333–339 (in Portuguese)

USDA (2005) 'Brazil: 2005/06 soybean area projected to decline', Report by the Production Estimates and Crop Assessments Division, Foreign Agricultural Service, United States Department of Agriculture, www.fas.usda.gov/pecad, accessed 5 April 2007

US-EPA (2006) *Global Anthropogenic Non-CO_2 Emissions: 1990–2020*, US Environmental Protection Agency, Washington, DC, USA

Valpassos, M., Cavalcante, E., Cassiolato, A. and Alves, M. (2001) 'Effects of soil management systems on soil microbial activity, bulk density and chemical properties', *Pesquisa Agropecuária Brasileira*, vol 36, no 12, online edition, www.scielo.br/scielo.php?script= sci_pdf&pid=S0100204X2001001200011&lng=en&nrm=iso&tlng=, accessed 6 June 2007

Van Oost, K., Govers, G., de Alba, S. and Quine, T. (2006) 'Tillage erosion: A review of controlling factors and implications for soil quality', *Progress in Physical Geography*, vol 30, no 4, pp443–466

Van Oost, K., Quine, T., Govers, G., De Gryze, S., Six, J., Harden, J., Ritchie, J., McCarty, G., Heckrath, G., Kosmas, C., Giraldez, J., Marques da Silva, J. and Merckx, R. (2007) 'The impact of agricultural soil erosion on the global carbon cycle', *Science*, vol 318, pp626–629

Van Oost, K., Six, J., Govers, G., Quine, T. and De Gryze, S. (2008) Response to letter from Lal, R. and Pimentel, D., *Science*, vol 319, pp1040–1042

Van Zwieten, L., Singh, B., Joseph, S., Kimber, S., Cowie, A. and Chan, K. (2009) 'Biochar and emissions of non-CO_2 greenhouse gases from soil', in J. Lehmann and S. Joseph (eds) *Biochar for Environmental Management*, Earthscan, London, UK, pp227–250

Vandermeer, J. (1995) 'The ecological basis of alternative agriculture', *Annual Review of Ecology and Systematics*, vol 26, pp201–224

VCS (2008a) *Voluntary Carbon Standard: Tool for AFOLU Methodological Issues*, VCS, Washington, DC, USA

VCS (2008b) *VCS Guidance for Agriculture, Forestry and Other Land Use Projects*, VCS, Washington, DC, USA

Villanueva, C., Ibrahim, M., Casasola, F. and Sepúlveda, C. (2011) 'Ecological indexing as a tool for the payment of ecosystem services in agricultural landscapes: The experience of the GEF Silvopastoral Project in Costa Rica, Nicaragua and Colombia', in B. Rapidel, F. DeClerck, J. Le Coq and J. Beer (eds) *Ecosystem Services from Agriculture and Agroforestry*, Earthscan, London, UK, pp141–158

Viola, E. (2004) 'Brazil in the context of global governance politics', *Ambiente & Sociedade*, vol 7, pp27–46

Vlek, P., Rodríguez-Kuhl, G. and Sommer, R. (2004) 'Energy use and CO_2 production in tropical agriculture and means or strategies for reduction or mitigation', *Environment, Development and Sustainability*, vol 6, pp213–233

Wiepolski, L. (2006) 'In situ noninvasive soil carbon analysis: Sample size and geostatistical conclusions', in R. Lal, C. Cerri, M. Bernoux, J. Etchevers and E. Cerrie (eds) *Carbon Sequestration in Soils of Latin America*, Food Products Press, Binghampton, NY, USA, pp443–456

Wilding, P., Drees, L. and Nordt, L. (2001) 'Spatial variability: Enhancing the mean estimate of organic and inorganic carbon in a sampling unit', in R. Lal, J. Kimble, R. Follett and B. Stewart (eds) *Assessment Methods for Soil Carbon*, CRC Press, Boca Raton, USA, pp69–86

Williams, T., Powell J. and Fernández-Rivera, S. (1995) 'Manure availability in relation to sustainable food crop production in semi-arid West Africa: Evidence from Niger', *Quarterly Journal of International Agriculture*, vol 34, no 3, pp248–258

Wittwer, S. (1995) *Food, Climate and Carbon Dioxide*, Lewis, Boca Raton, USA

Woolf, D., Amonette, J., Street-Perrott, F., Lehmann, J. and Joseph, S. (2010) 'Sustainable biochar to mitigate global climate change', *Nature Communications*, vol 1, article 56, doi:10.1038/ncomms1053

Woomer, P., Karanja, N. and Murage, E. (2001) 'Estimating total system C in smallhold farming systems of the East African highlands', in R. Lal, J. Kimble, R. Follett and

B. Stewart B (eds) *Assessment Methods for Soil Carbon*, CRC Press, Boca Raton, USA, pp147–166

World Bank (2008) *Implementation and Completion Report on Integrated Silvipastoral Approaches to Ecosystem Management Project in Colombia, Costa Rica, and Nicaragua*, The World Bank, Washington, DC, USA

WRI (1990) *World Resources 1990–91*, Oxford University Press, New York, USA/Oxford, United Kingdom

Index